# Archaeological Studies of Pacific Stone Resources

Edited by

Foss Leach and Janet Davidson

BAR International Series 104
1981

# B.A.R.

B.A.R., 122 Banbury Road, Oxford OX2 7BP, England





B.A.R. S104, 1981, "Archaeological Studies of Pacific Stone Resources"





ISBN 9780860541226 paperback

ISBN 9781407324388 e-book

DOI https://doi.org/10.30861/9780860541226

A catalogue record for this book is available from the British Library

This book is available at www.barpublishing.com

CONTENTS

# PREFACE

The papers in this volume were offered at a symposium on 'Lithic Resources in the Pacific Region' at the XIV Pacific Science Congress in Khabarovsk, USSR, in August 1979. For various reasons, not all authors were able to attend the symposium in person, and several papers were read by colleagues for them.

The symposium was organized to bring together people working on lithic materials from archaeological sites in the Pacific Region, particularly those interested in prehistoric movements of raw materials associated with economic and ceremonial exchange networks. Valuable raw materials such as stone have moved from one part of the Pacific to another, either with founder populations seeking new territory, or as part of regular communication patterns between one group and another. The archaeologist wishing to identify and describe this process has been drawn deeply into several branches of the physical sciences in order to determine accurately the original geographic sources of stone tools, and also to date their manufacture. This has meant a great deal of cooperative research between archaeologists and researchers in the chemical sciences, geology and nuclear physics particularly. This spirit of cooperation has been very fruitful, and has lead to greater discipline in archaeological interpretation. The papers in this volume illustrate this concern to refine the tools and methods of prehistoric research.

The participants of the symposium passed a number of resolutions as follows:

1. That in order to promote the study of prehistoric lithic resources in the Pacific, there should be a continuing exchange of publications and information between participants in the symposium and other interested colleagues.

2. That the proceedings of the symposium should be made known as soon as possible to colleagues who were not able to be present, by the prompt publication of summaries in various journals.

3. That the papers of the symposium should be published in full in a single volume.

4. That the development of characterization studies of all types of rock of economic importance to prehistoric people in the Pacific and its rim should be strongly encouraged.

5. That an archive should be established at the Australian National University to include samples from all known obsidian sources in the Pacific and its rim; and that this archive should be freely available for use by researchers.

6. That in future, element composition studies of Pacific obsidians should refer to two standard sources: one for the Oceanic obsidian types (that from Mayor Island in New Zealand), and another for non-Oceanic types.

Foss Leach and
Janet Davidson (Editors)
May 1980

## LIST OF CONTRIBUTORS

W. R. Ambrose,
Department of Prehistory,
Research School of Pacific Studies,
Australian National University,
Box 4 PO,
Canberra ACT,
Australia 2600.

Russel J. Beck,
Southland Museum and Art Gallery,
PO Box 1012,
Invercargill,
New Zealand.

J. R. Bird,
Australian Atomic Energy Commission Research Establishment,
Lucas Heights,
Private Bag Sutherland,
NSW 2232,
Australia.

T. P. Butylina,
Pacific Institute of Geography,
Far East Science Centre,
USSR Academy of Sciences,
17 Uborevicha St.,
Vladivostock,
USSR 690000.

Т.П.БУТЫЛИНА,
Тихоокеанский институт географии
ДВНЦ АН СССР Владивосток, СССР.

Hamish J. Campbell,
NZ Geological Survey,
PO Box 30368,
Lower Hutt,
New Zealand.

P. J. F. Coutts,
Victoria Archaeological Survey,
29-31 Victoria Avenue,
Albert Park,
Melbourne,
Victoria 3206,
Australia.

Janet Davidson,
Anthropology Department,
University of Otago,
PO Box 56,
Dunedin,
New Zealand.

P. Duerden,
Australian Atomic Energy Commission Research Establishment,
Lucas Heights,
Private Bag Sutherland,
NSW 2232,
Australia.

Graham Harrison,
Department of Prehistory,
School of General Studies,
Australian National University,
Box 4 PO,
Canberra ACT,
Australia 2600.

Takenobu Higashimura,
Research Reactor Institute,
Kyoto University,
Kumatori-cho,
Sennan-gun,
Osaka-fu,
Japan.

Yoshimasa Kamaki,
Okayama College of Science,
Ridai-cho,
Okayama,
Japan.

G. M. Kovnurko,
Geological Division,
Leningrad State University,
University naberezhnaja 7-9,
Leningrad,
USSR 199164.

Г.М.КОВНУРКО,
Ленинградский Государственный
Университет им.А.А.Жданова,СССР.

B. Foss Leach,
Anthropology Department,
University of Otago,
PO Box 56, Dunedin,
New Zealand.

Helen M. Leach,
Anthropology Department,
University of Otago,
PO Box 56,
Dunedin,
New Zealand.

Isabel McBryde,
Department of Prehistory,
School of General Studies,
Australian National University,
Box 4 PO,
Canberra ACT,
Australia 2600.

V. I. Mishin,
North-Western Territorial Geological Office,
USSR,

В.И.МИШИН,
Северо-Западное Территориальное
Геологическое Управление, СССР.

G. I. Suslov,
North-Western Territorial Geological Office,
USSR.

Г.И.СУСЛОВ,
Северо-Западное Территориальное
Геологическое Управление, СССР.

Douglas G. Sutton,
Anthropology Department,
National Museum of Natural History,
Smithsonian Institution,
Washington DC,
USA 20560.

Tetsuo Warashina,
Research Reactor Institute,
Kyoto University,
Kumatori-cho,
Sennan-gun,
Osaka-fu,
Japan.

S. E. Warren,
Physics Department,
University of Bradford,
Bradford BD7-1DP,
United Kingdom.

J. P. Wesson,
Victoria Archaeological Survey,
29-31 Victoria Avenue,
Albert Park,
Melbourne,
Victoria 3206,
Australia.

Tigger Wise,
58 Cambridge Avenue,
Vaucluse,
NSW 2030,
Australia.

# 1. THE IMPERMANENCE OF OBSIDIAN SOURCES IN MELANESIA

W. R. Ambrose, J. R. Bird and P. Duerden

The circumpacific region is characterized geologically by active volcanicity, one product of which is the volcanic glass obsidian which was an important resource for prehistoric communities. The island arcs of Melanesia are volcanically active and in four widely separated areas obsidian or pitchstone was available and exploited prehistorically. Archaeologists have for some time appreciated the value of obsidian to prehistoric non-metal using communities and have therefore devised ways for gleaning as much information as possible on the use and distribution of this resource.

The particular value of obsidian to archaeological studies arises from the following considerations:

1. The material was sought after by prehistoric communities because of its razor sharp edge which was suited to fine cutting operations. A systematic study of flake technology and use wear has not been made for Melanesian artefacts in obsidian; but because of its fragility, sensitivity and therefore the response of the cutting edge to different cutting operations, a technological study of microscopic use-wear could be useful.

2. Obsidian is relatively rare in occurrence, being confined to localized flows. In Melanesia only four general areas have yielded artefact quality obsidian. This has meant that obsidian was purposefully mined for later distribution over long distances. The material therefore provides a good base for what Renfrew (1969: 157) has termed 'trade statistics'.

3. Obsidian is very durable in archaeological deposits and can remain apparently unweathered for thousands of years. Its durability, as a siliceous material, is probably second only to quartz. The weathering of obsidian is basically an hydration process and under favourable conditions the hydration rate can be calculated to produce an estimated time for the production of a flake surface.

4. The cutting edge of an obsidian sherd is brittle and it therefore has a short life as a cutting implement. The problem of hoarding and constant reuse which may be characteristic of other stone tools in Melanesia (White and Modjeska, 1978), and which may cause problems of anachronisms or distorted distributions, is therefore reduced. As artefacts of the area in which they are found, small useless but used flakes of obsidian should therefore be fairly reliable indicators of activity areas. Distributional data should therefore be directly reconciled with a simple use and discard pattern.

5. Most flows of obsidian were geologically instantaneous events with a high degree of chemical homogeneity frozen into their composition. Post

depositional weathering affects only the obsidian surface, while deep weathering which converts it to crumbly perlite, would make the material useless for artefacts. Therefore, the vitreous unweathered obsidian which would be used for artefacts should be chemically representative of the composition of its parent flow at its time of formation.

6. The homogeneity of an individual flow means that even very small artefacts or flakes, down to a millimetre or so, can be chemically analysed with the high probability that the results will be consistent with the composition of the parent flow.

Some of the foregoing observations rely on the explicit demonstration of shared chemical identify between an artefact and a source. The assurance with which this identification can be made is very dependent on the quality and quantity of the analyses. If for instance chemical data was available only for the two major obsidian forming elements Silicon and Aluminium there would be a poor prospect of clearly differentiating between sources in the same geological province and even less chance of differentiating between flows in a single volcanic complex. Fortunately, obsidian is ideally suited to some techniques of chemical analyses because of the inherent variability in its trace element composition which reflects the variability between individual volcanic eruptions.

MELANESIAN SOURCES

The first necessity in sourcing studies is of course to find the sources. Geological surveys in Melanesian volcanic areas which might produce glassy products are fairly recent. The first comprehensive record of obsidians from a geological point of view was reported in 1974 (Smith), where the chemical analyses and the field sampling procedures were primarily designed for geological purposes. Many of the analysed obsidians could be described as glassy rock which no amount of knapping skill could convert to useful cutting tools. A more specifically archaeological sources survey was made in 1974 (Ambrose, 1976a) of all the areas where obsidian was geologically reported. Samples were gathered for an archaeologically oriented series of analyses which were carried out at the Australian Atomic Energy Commission using Neutron activation (Wall, 1976) and Prompt Nuclear analysis (Bird and Russell, 1976). Technical advances, particularly in the prompt nuclear analytical techniques of proton induced gamma emission (PIGME), and proton induced X-ray emission (PIXE), have provided archaeologists with more than 2000 chemical determinations for obsidians from Melanesian sites over the last couple of years. Subsequent statistical analysis of more than 1000 obsidian flakes from various archaeological sites pointed toward a rather fundamental problem in adopting the normal and simple allocation of artefacts to their source material based on chemical similarity. The precision in the analytical technique and the large number of analyses which were performed showed clear discrepancies in the 'fit' of some artefacts to source. It appears that more diverse materials were used prehistorically than it is now possible to locate using normal field sampling techniques in the Melanesian region.

The four main source regions which could geologically produce obsidian are well known as:

1. The Lou and Pam islands to the south of Manus Island in the Admiralty Islands.
2. The area of the Willaumez Peninsula and Cape Hoskins near Talasea on the north coast of New Britain.
3. Fergusson, Sanaroa and Dobu islands in the D'Entrecasteaux Islands.
4. Gaua and Vanua Lava Islands in the Banks Group of the New Hebrides island chain.

All of these regions contain multiple utilizable obsidian deposits not all of which appear to be present in the analysed archaeological collections from sites elsewhere, and conversely where, as mentioned above, not all the prehistorically used artefact obsidians are recongizable as visible sources.

THE SAMPLING PROBLEM

All four geological areas in Melanesia which have produced obsidians are marked by active volcanicity which has continued from the Pleistocene to the present. The problem of sampling from these sources can be appreciated from the following outline of the geological instability of each source area.

A. Lou Island comprises 12 coalesced volcanoes which have produced pumice, ash and some flow deposits (Johnson and Smith, 1974: 334). There are solfataric areas with hot springs and mud pools which indicate the recent age of some of the volcanic deposits. A soil horizon, buried beneath 3 m of pumice and ash, near the present Rei village, contains pottery and obsidian. The 3 m ash and pumice deposit would be cataclysmic for any settlements and the vegetation for most of this 12 km long island. Recent work on the volcanic history of Lou (Paine, 1980) has shown that at least two other similar cataclysmic eruptions have occurred before 2000 years ago. The most recent volcanic event was in 1953 when Tuluman Island emerged from the sea at the southern tip of Lou. It now stands as a smaller island than previously (Reynolds and Best, 1976) but has large pitchstone-obsidian cliffs. In addition to the island building eruptions which are obvious, there is also evidence for land subsidence on some south coastal areas.

B. The Talasea-Cape Hoskins area of New Britain is a region of fairly massive volcanic flows and air-fall deposits. The Willaumez Peninsula comprises a chain of 11 Quaternary volcanoes (Lowder and Carmichael, 1970: 18) while there are widespread hot springs and solfataric areas at the present time (Johnson and Blake, 1972: 55). At the northern end of the peninsula both Mount Makalia and Mount Pago erupted in the late nineteenth century. The Cape Hoskins volcanoes 50 km to the east have produced mountainous deposits of pumice and ash over the last two and a half thousand years, some of these burying ancient soil horizons (Blake and McDougall, 1973: 202). One of these soil horizons, at Waisisi, contains obsidian flake tools with a chemical composition different from any presently known source. Johnson and Blake (1972: 33) believe that some of the Cape Hoskins volcanic tephra has been contributed as air-borne ash from the Willaumez volcanoes 50 km away. There are important implications in these events for the permanent disruption in the supply of prehistoric obsidian and the widespread dislocation of settlements in the region. In the Talasea area there are archaeological deposits which have been sealed by volcanic ashes. There is also some evidence for land subsidence with occupation debris in the form of obsidian flakes and 2000 to 3000 year old pottery being exposed on mud flats at low tide around Talasea (Specht, 1974: 303). In contrast Lowder and Carmichael (1970: 23) reported raised coral reefs in the vicinity. Clearly, volcano building and tectonic land movement have played havoc with the face of the landscape over the last few thousand years. Yet Talasea obsidian has been exploited for at least the last 6000 years (White et al., 1978:877; Ambrose, 1976a: 367). It is unlikely that the individual flows in this region which provided changing sources of obsidian over this period will all be located. The buried Waisisi artefacts near Cape Hoskins are an example of material from a local but unknown source which may never be found beneath the mountains of later volcanic debris. The Waisisi 'source', from the buried soil dated to 2590±250 (GaK-3076) (Blake and McDougall, 1973: 202), on the south of Mount Oto is most economically explained as being from a local deposit; Blake and Ewart (1974: 320) refer to fragments of welded tuff with rhyolitic composition from Witori volcano at Cape Hoskins.

C. The D'Entrecasteaux Islands sources on Fergusson, Sanaroa and Dobu Islands are the result of volcanic events within Pleistocene to Recent ages. Judging from the contents of archaeological sites on the Papua New Guinea mainland, the most important source of obsidian was the Kukuia peninsula on the western end of Fergusson Island. The peninsula sources appear to be in a relatively stable area with Potassium-Argon dates between 0.41 to 1.3 million years for its formation (Smith, 1973: 22). Nearby, another western Fergusson source at Fagululu is considered to be the most active thermal area in eastern Papua New Guinea and is considered to be very recent (ibid.: 8). Eastern Fergusson Island is a very active volcanic zone also producing usable obsidians. The Aiasuna village deposit, on the south side of Numanuma Bay is at the eastern flank of Mount Oiau which has been marked by explosive pumice ejection within the last 100 to 200 years (ibid.: 12).

Adjacent Dobu Island is considered to be an active volcano, temporarily quiescent, which is composed of recent pumiceous ashes, rhyolite and obsidian flows. Events which produced one of these major flows occurred within the last two centuries (ibid.: 11).

Another geographically associated obsidian source is on Sanaroa Island off the eastern end of Fergusson Island. Sanaroa is mainly composed of basaltic volcanic rocks with some rhyolite and obsidian, the latter occurring in small pieces which appear to have been thrown out as explosion debris. The glass is nevertheless of very high quality. The age of the Sanaroa volcanics is not known but judging by the lack of youthful volcanic landforms (ibid.: 14), it is likely that the obsidian from here has been available for several thousand years. However, no Sanaroa obsidian has so far been recognized in analyses of archaeological collections of several hundred flakes from Papua New Guinea sites.

D. The Banks Islands glasses are from Gaua and Vanua Lava Islands which are youthful volcanoes. Some activity is persisting in geologically related volcanoes up to the present (Mallick, 1973: 196). The flake quality of these glasses, which are probably better described as pitchstones, is sufficiently good to produce usable flakes. But their quality does not match that of the Papua New Guinea sources.

## DIFFICULTIES IN SAMPLING

The Melanesian obsidian source areas present difficulties which are not normally encountered in archaeological sourcing projects. In particular, the three Papua New Guinea source areas have been subject to cataclysmic volcanic events over a time range spanning human occupation in these islands. Nevertheless it is likely that reliable allocation of artefacts to their sources is possible in instances where a source has been continuously available. On the other hand, two other possibilities have to be countenanced: that some sources which were once available to prehistoric communities have been obliterated; and, additional more recent sources have become available since human settlement in the areas concerned.

The ideal strategy under these difficult conditions should therefore encompass the conventional approach of allocating artefacts to known sources, and include the search for unknown source material in the archaeological collections. It is important to acknowledge that the presently known sources do not represent the total amount of chemical variability which has existed in obsidian sources over the prehistoric time span of the region. In addition, this chemical variability of obsidians in archaeological collections is likely to be somewhat different over time, and reflect the changingavailability of raw material. With this in mind a detailed analysis on the Admiralty Islands obsidian has been undertaken as part of a general enquiry into the prehistory of this area.

## THE ADMIRALTY ISLANDS OBSIDIANS

There are three Admiralty islands which contain obsidian deposits and all have been exploited in prehistoric times. They are Pam Lin, Pam Mandian, and Lou Island in the St Andrew Strait off the south coast of Manus Island (Figure 1.1). Previous work on the characterization of locatable sources has been reported (Bird and Russell, 1976; Wall, 1976; Ambrose, 1976a; 1978; Bird, et al., 1978; Duerden et al., 1979; 1980).

The volcanoes comprising the central axis of Lou Island have generally grown in sequence toward the south-west (Johnson and Smith, 1974: 335). The most recent is the offshore islet of Tuluman which emerged in 1954. However, some vents of these Lou Island volcanoes have been reactivated over time; for instance, the eastern volcanic centre, though thought to be the earliest formed, was last active around 2000 years ago. The evidence for this comes from excavations in Rei village where a 3 m ash and pumice deposit overlay and obliterated on obsidian working floor where scores of triangular-sectioned points and thousands of waste flakes, along with hearths and charcoal were abruptly blanketed. Unfortunately there are no massive obsidian flows visible as part of this later dated eruption or any other earlier undated volcanic events on Lou Island. The precise location of the source for the 2000 year old Rei artefacts is not known.

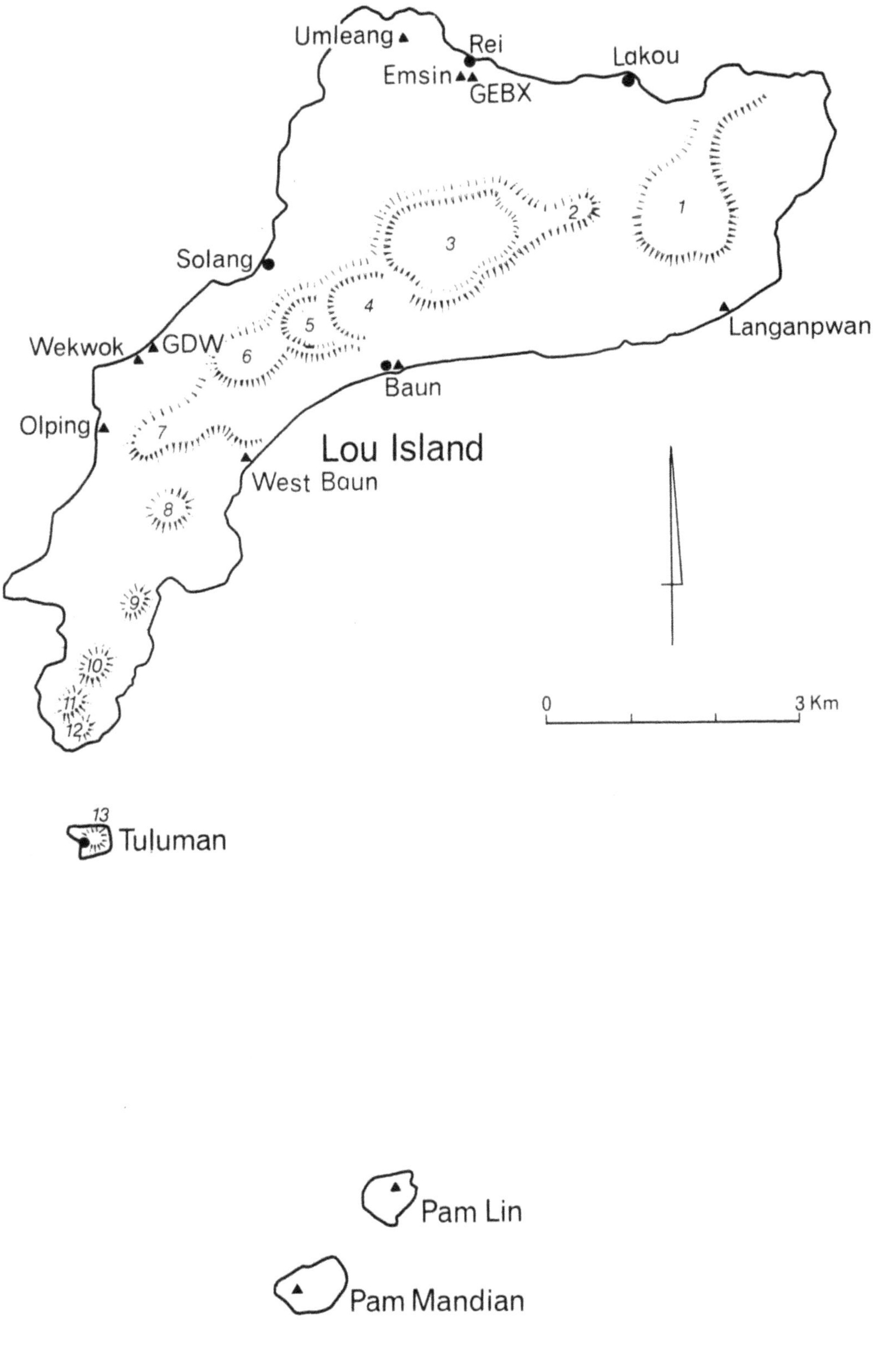


Figure 1.1 Obsidian sources in the Admiralty Islands

About one kilometre west of Rei village are at least 25 quarrying shafts, with one 17 m deep, which penetrate the overlying 2000 year old 3 m ash deposits in order to recover large lumps of obsidian embedded in underlying ash beds. This instance is the only one at present known for quarrying operations on Lou. It was first reported by Parkinson (1907: 373) who visited the area at the turn of the century. Present day informants at Rei claimed that similar mine shafts were dug west of Solang village about 5 km southwest of Rei but the site has not been located. Circumstantial evidence for the existence of the second quarry area is the presence of an obsidian working floor, at Wekwok, with extensive discard flakes, cores and some flake implements. Lou Island is therefore not an ideal situation in which to carry out direct sampling of obsidians if the aim is to differentiate between individual flows. This shortcoming is partly due to the mode of occurrence of the obsidian as boulder and cobble-sized lumps in pumice ash breccias, but mainly due to the changing volcanic landscape over time.

The two Pam islets appear not to have suffered the same effects. Pam Lin, which is about 500 m in diameter, has a centrally located obsidian deposit which has an elevation of only about 2 m above high water mark. Its relative eustatic stability is not known. Pam Mandian is an islet which has boulders and flows of rhyolite-obsidian on its west coast. The obsidian is of poor quality but it can produce useful implements and these have been recognized in artefact collections from elsewhere.

## SAMPLING STRATEGY

As it is not possible in all cases to sample from primary geological deposits these have been supplemented by archaeologically assembled artefacts, from waste flakes and spear points, surface collections and stratified deposits.

Since the only obsidians available in the Admiralty Islands are from restricted deposits on the St Andrew Strait Islands of Lou and Pam, it is reasonable to adopt the view that different artefact collections from the source islands will fairly represent the materials being exploited on those islands. The inventory of the Lou-Pam artefact obsidians will therefore be taken as a fair reflection of the sources, their relative importance, and their change over time. Information on the chronology of obsidian source exploitation may therefore be applied to sites elsewhere on Manus and the non-obsidian producing islands of the archipelago. Lou Island exploitation chronology can also be applied to sites far distant from where the obsidian has been transported in prehistoric times.

Surface collections were made on the Pam Islands of likely artefact quality obsidians. Road and beach exposures of archaeological deposits on Lou Island were sampled for obsidian flakes, and two excavations were carried out—one at the quarry shaft site at Umleang, and the other at Emsin's site in Rei village —to sample prehistoric strata for their obsidian contents. Twenty-eight groups of obsidian were assembled in this procedure comprising nine surface collections, twelve stratigraphic level collections from the shaft quarry site, and seven stratigraphic collections from the Rei village site. About 300 specimens were then submitted from these collections for chemical analysis. The chemical and statistical analysis of these specimens was intended to throw light on the chronology and provenance of surface and excavated collections from elsewhere in the Admiralty Islands. Thus, in total, about 750 specimens were chemically characterized in order to define the source groups and allocate other collected artefacts to these. This approach to the sourcing of obsidian would not be feasible without the availability of a fast and efficient chemical analytical technique, and computer based multivariate analysis.

PIXE ANALYSIS

Recent advances in proton induced X-ray emission analysis and its application to glasses (Duerden et al., 1979) has opened new possibilities for the examination of obsidians which in turn allows questions of prehistoric communication networks to be investigated in a more quantitative way.

The analytical technique was developed to accommodate a wide range of obsidian flake sizes and to enable solid, unprepared specimens to be processed. Conventional analytical techniques such as X-ray fluorescence and Atomic Absorption analysis consume the artefact, or part of it, in sample preparation. Non-destructive activation analysis may leave residual radioactivity in the specimen or suffer poor reproducibility with solid samples, because of variable specimen shape.

In the case of proton induced X-ray and gamma-ray emission analysis (PIXE-PIGME) solid specimens down to 5 mm in diameter were acceptable provided that a reasonably flat area of about 3 mm diameter could be located within the focus of the proton beam. No residual activity is induced in the specimen, so that no special handling procedures were required. Up to 60 specimens, including repetitions of standard glasses, were loaded for each analysis cycle. Each specimen was irradiated for about 4 minutes after which the accumulated data was dumped into the computer storage files and the next sample was automatically presented to the beam. In this fashion 60 specimens could be analysed in about four hours.

Simultaneous Gamma ray and X-ray spectra were recorded. Counts in the Na 440 keV and 1630 keV peaks, the Al 843 keV and 1013 keV peaks and in the region 5 to 7.2 meV are derived from the $\gamma$-ray spectra and have been converted to absolute concentrations for sodium, aluminium and fluorine respectively. From the X-ray data, 19 energy intervals were selected to show the behaviour of prominent peaks in the X-ray spectrum. Total counts for each interval give information on Si, Cl, K, Ca Ti, Mn, Fe, Cu, Zn±Cu, Ga, Zn, As+Pb, Rb, Sr, Y+Fe+Rb, Zr+Sr, Nb+Y, Zr, and Nb X-rays. Absolute concentrations of elements can be determined (Duerden et al., 1979) but require lengthy computer unfolding of the X-ray spectra. Since there is broad geochemical homogeneity in the quantitatively analysed Lou-Pam obsidians (Johnson et al., 1978: 57) it is felt that an analysis based on the X-ray counts together with the 3 $\gamma$-ray concentrations is sufficient for this study.

Pixe analysis has been used on prepared obsidians with useful results (Neilson et al., 1976). The present technique developed at the Australian Atomic Energy Commission is therefore even more attractive for archaeological materials because of its ability to utilize solid and small specimens down to 5 mm in diameter. The rapidity and minimum preparation necessary for the artefacts to be analysed has made it possible to produce analytical data in quantities sufficient to explore archaeological questions which were not possible previously. The statistical analysis of the large amount of generated data begins at its time of generation.

Pulses from the X-ray and gamma-ray detectors are recorded on an on-line computer (PDP-15) and periodically archived to a larger unit. The computer also controls the mechanical functions of the sample carrier by activating timing and sample changing operations (Duerden et al., 1980).

DATA ANALYSIS

The large data store of more than 2000 chemical analyses in up to 19 X-ray energy windows, and three gamma-ray concentrations can only be practicably handled by computer using multivariate statistical techniques which can use all the data simultaneously to the best advantage.

Multivariate data analysis in archaeological applications has been reviewed by Doran and Hodson (1975) who discuss the relative merits of a range of procedures, from automatic classification to the discrimination and characterization of known groups in classification procedures.

The procedure in the detailed analysis of the Lou Island obsidians was to use the archaeological information of both stratigraphic, and therefore chronologically separate collections of obsidian, and geographically separate, collections of unknown age as the starting points, or preliminary working clusters, for statistically defining groups. It was assumed that the total variation in chemical composition of obsidians in the Admiralty Islands is represented in the universe of the combined 28 Lou groups. Each group collection was treated as if it was representative, with equal probability, of a larger parent population. It was expected that this unlikely working presumption would be progressively proved by the developing classification which is described below.

Some of the 28 initial collection groups were presumed to have no separable identity, either because they were samples from the same parent group and shared the same multivariate properties as another collection, or because they were mixed collections composed of specimens from more than one parent population.

Two procedures were used to reduce the number of groups to what appeared to be a minimum stable number of groups.

A. Automatic Classification

The obsidian data is quite suitable for analysis by techniques which combine the total multivariate information to generate groups and subgroups using various sorting and clustering procedures. The methods included in Wishart's (1971) CLUSTAN manual achieve group clusters from unclassified cases in a variety of statistical ways. These are referred to as general strategies of (a) hierarchic fusion, with ten different clustering criteria; (b) monothetic division, with four possible critera; (c) iterative relocation with five clustering criteria and (d) three other methods. Iterative relocation, using the squared euclidean distance similarity coefficient, was adopted in this analysis.

The CLUSTAN programme 'RELOC' has already been applied to archaeological data (Ottoway, 1974) for an analysis of British stone tools, where use was made of the 'dissimilarity coefficient' to evaluate the statistical distance of an artefact from the closest group centroid. RELOC requires a starting set of either randomly chosen or prior clusters. The data is scanned for each case's similarity to the specified groups. The programme therefore relocates any case to its nearest cluster and recomputes the changed group centroid on each occasion. When cluster stability is achieved after a specified number of iterative relocations of cases, a similarity matrix between all cluster pairs is produced. The pair of clusters showing the smallest dissimilarity coefficient are fused and the next relocation cycle commences. The iterative relocation of cases and hierarchic fusion of clusters proceeds until the specified minimum useful groups are achieved.

In order to reduce computing time, an arbitrary reduction of groups to 18 was made by combining several of the stratigraphically adjacent collections from the Umleang and Emsin deposits.

The original numeric data was standardized and a principal component analysis performed on this data. Ten principal component scores were used in the subsequent analysis using RELOC.

In the first pass through the data of the entire source group of 338 cases there were 217 relocations and the cluster groups were reduced from 18 to 16. In the reduction of the entire group from 18 to 9 groups there were 747 case relocations during 21 iterations. This appears to indicate flabbiness in some of the initial 18 groups.

Nine groups were chosen as the final stable clusters (Figure 1.2) because a maximum amount of archaeological information appears to be present consistent with a minimum number of groups. For example reductions to eight groups would have combined clusters 1 and 4 which would obscure a differentiation, between the Emsin and Umleang sites, that is warranted on stratigraphic grounds. Further reduction to seven groups would combine clusters 6 and 7 which again would obscure the difference between the lower and middle levels of the Emsin site. An increase in group numbers beyond nine tends to split what appear to be related clusters. For example 10 groups would require the splitting of cluster 7 into two groups which roughly share the same stratigraphic distribution. When the 'ten-group' solution was applied to the collection of obsidians from the excavations at Kohin cave, on the Manus mainland, a similarly parallel stratigraphic association occurred. For these reasons nine groups has been adopted as a satisfactory minimum set of stable clusters.

The original group association of the final clusters, whether geographic or stratigraphic, has been severed to a variable extent by the relocation and iteration cycles of the RELOC programme. The unequal effect can be seen in the case of the initial group collections numbered 4, 7, 8, 9, 15, 21 and 27, which are practically unsorted compared with other groups such as those numbered 1, 3, 11 (Figure 1.2) which are split between several groups.

B. Classification from prior groups

In order to classify the many hundreds of non 'source' collections of Admiralty Islands obsidian it was necessary firstly to define the clusters which

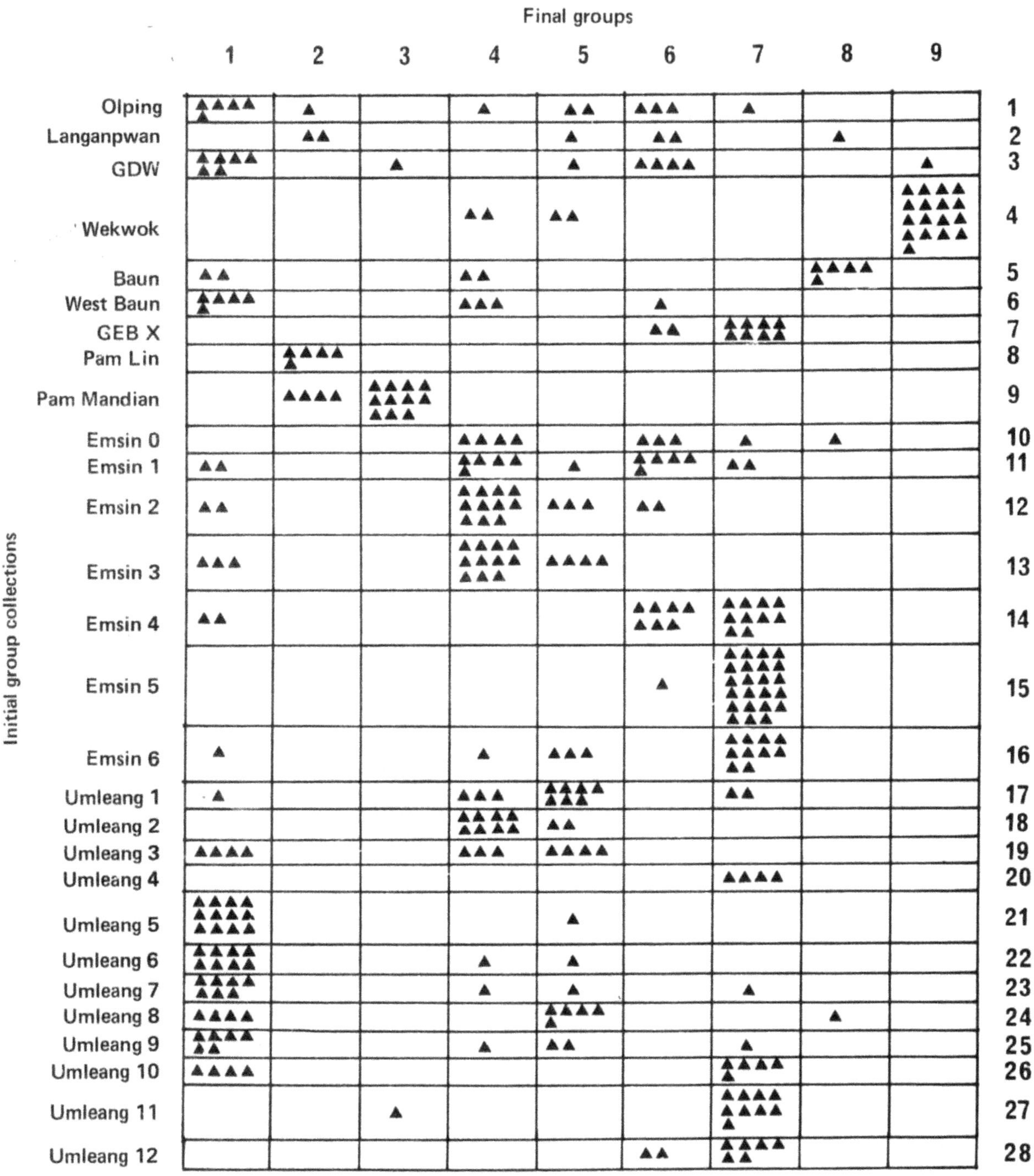


Figure 1.2 The nine final obsidian 'source' groups

had been extracted by the RELOC programme. The multivariate discriminant function analysis described by Klecka (1975: 434) was used for this task. The discriminant programme reduced the 22 variables from the X-ray and gamma-ray analysis to eight discriminant functions and determined the 9 group centroids and dispersions of case data with respect to the centroids. Inspection of the multivariate F matrix of group pairs gave a scaled indication of the similarity between all group pairs and the significance of their association. This matrix consists of an F ratio for each pair of groups, and is the significance test for the Mahalanobis distance between groups. With an assumption of random sampling and multivariate normal distributions, the F value can be used to test for the equality of pairs of centroids (ibid.: 460). The two groups showing the closest affinity were 6 and 7, which were also the groups the CLUSTAN programme would combine as the closest pair, as mentioned previously. The probability of each case being part of the population of the group it was allocated to was also calculated. In this analysis, of the cases which were allocated to one of the nine groups in the CLUSTAN programme, 96.1% were similarly allocated in the Mahalanobis discriminant programme.

ALLOCATION OF ARTEFACTS TO PRIOR GROUPS

The data from the 338 cases of known stratigraphic and surface groups, having been subject to the CLUSTAN procedures were processed as a final set of 'source' groups. The separation between these nine clusters, which is maximized by the Mahalanobis $D^2$ analysis, required the computation of eight canonical discriminant functions to account for 100% of the between-group variance. The need for so many functions is an indication of chemical closeness of some of the groups. In a previous analysis of Melanesian-wide obsidian source groups most between-group variance could be accounted for with as few as three canonical discriminant functions, and this is some indication of much larger chemical differences at the broad regional level than exists within the narrow geochemical provence of the Lou-Pam obisidians in the Admiralty Islands.

The cluster centroids, defined by the discriminant functions as positions in multidimensional space, were then used in the reallocation phase of the analysis, to find the closest centroid to each of the 338 cases in turn. Since in this operation 96% of the cases were allocated to their prior group, the clusters were considered to be sufficiently well separated to be used as 'groups' in a classification of other Admiralty Islands obsidians.

The chemical analysis data from 490 ungrouped obsidians from surface sites in the Bismarck Archipelago were processed at the classification stage of the discriminant function analysis. The programme, having reduced the chemical analyses on the 2 element $\gamma$- and X-ray counts to eight discriminant functions, then attributes each case to the closest group. An index of distance from the group centroid, and the probability of group membership, for each case is calculated. It is therefore possible to set limiting values for acceptance or rejection of group membership. In Figure 1.4 and 1.5, cases which are plotted as open triangles are identified with particular groups with a low probability of their ascription. This raises a problem which has not been fully investigated relating to the surface chemical alteration of obsidians in some aggressive soil environments (Ambrose, 1976b: 84). Kohin Cave obsidians

(Figure 1.4) show some cases where the normal glassy flake surface has been altered by chemical erosion to appear dull and matte. The low probability of cases at Kohin and other Bismarck Archipelago sites is likely to be a reflection of selective chemical alteration, but the possibility also exists that the cases in question are exhibiting a range of chemical variation which is not equalled in the nine 'source' groups. That is, there may be evidence for other unknown sources. Despite these reservations the results are very envouraging and further analysis of unknown source obsidians from other Bismarck Archipelago collections is in progress. The results presented here are therefore incomplete, there being only 490 unknown source specimens so far analysed. A further thousand analyses would be necessary to cover the archaeological collections from the region adequately.

A few indications of what the analysis has achieved may show the value of the analysis of large numbers of artefacts using the systems outlined before.

Figure 1.1 is a map of the source areas in the Lou-Pam Islands. The 13 main volcanic centres are numbered in italics. Vent 13, Tuluman Island, emerged from the sea in 1954 and although a low flake quality obsidian was extruded at that time it has not been included in the present analysis. The chemical composition of the Tuluman obsidian is nevertheless distinguishable from the other Lou-Pam sources. The triangles in the map indicate areas from which collections for the 338 analysed 'source' specimens were made.

The final groups are presented in Figure 1.2. It can be seen that some of the initial group collections are still intact while others have undergone considerable reshuffling.

Group 1 dominates throughout the use of the Umleang site. The deposit has been laid down on a soil developed on Rei tephra (Pain, 1980) which has been carbon dated at between 1700 and 2000 years ago (ANU 2018, 1710± 70 BP; ANU 2193, 1450±100 BP; ANU 2194, 2100±160 BP). Layer 4 at Umleang has been dated at 220±80 BP (ANU 2019).

Group 2 is the Pam Lin quarry source from which specimens of waste material were recovered.

Group 3 is from surface collections from the southern half of Pam Mandian island and includes natural source material as well as flaked artefacts.

Group 4 appears to be the dominant source for the upper levels of the Emsin site which is buried beneath the Rei tephra and is dated to 2100±160 BP (ANU 2194). The same source is contributing to the upper levels of the Umleang site which is less than 220 BP. The apparent inconsistency is explained by the fact that levels 1-4 at Umleang are related to obsidian mining by way of deep shafts to material in ash deposits beneath the Rei ash. It therefore is apparent that the very recent mining is quarrying obsidian which is the same as, or chemically indistinguishable from, the earlier source for the Emsin site.

Group 5 has a similar distribution to group 4 and is predominantly an Umleang associated obsidian.

Group 6 appears to have been a relatively early source being mainly associated with the 2000 year old Emsin deposit.

Group 7 has an early association with both the pre-Rei ash at Emsin site and the post-Rei soil at the base of the Umleang site.

Group 8 is a small collection which is unique to the Baun area. It does not appear to have been widely utilized as only one archaeological specimen has so far been located, that is at the Lesu site on the New Ireland north coast (Figure 1.5).

Group 9 is a unique discrete collection from a working floor at the west end of Lou Island. Like group 8, its lack of wide distribution in other Bismarck Archipelago sites is a distinctive feature compared with groups 1.7.

Figure 1.3 presents the information on the derivation of the obsidian excavated at the Ahus Island site. Ahus is a small, 500 m by 1000 m, coral island 3.5 km from the mid-north coast of Manus Island. The excavation was made in a low 0.5 m mound which was expected to yield a basal date of less than 200 years, based on traditional information for the first settlement of the island. The radiocarbon date for lower levels in fact is 'modern' or in other words less than around 220 years. The dominance of the group 1 source for all levels apart from the top is in contrast to the very late levels at Umleang (Figure 1.2) but is consistent with Umleang Layer 5 which has been dated to around 220±80.

Figure 1.4 shows the obsidian distribution for the Kohin Cave site (Kennedy, 1980). The surface of many specimens shows etch pitting caused by the aggressive damp conditions in the cave floor. This is likely to be the reason that many specimens are attributed to a group source with a low degree of probability, and these are plotted as open triangles. The most distinctive feature of this distribution is the presence of Pam Island sources at the earliest levels in the site which have been dated to around 3450 BP (ibid.).

Figure 1.5 refers to three collections from Masahet Island in the Lihir group, Eloaue in the St Matthias group, and Lesu on the New Ireland mainland. Masahet is a small raised coral island with an encircling foot track connecting all the coastal villages. The analysed obsidians are from surface collections at Ton village. The dominance of 6-7 group could indicate a very recent pattern, as with the top level of Ahus Island site (Figure 1.3) or a period between 1000 to 2000 years ago when compared with Emsin stte (Figure 1.2) or Kohin Cave Layer 4 (Figure 1.4). Masahet is about 600 km straight line distance from the Lou Island sources. Eloaue, about 300 km from Lou Island, is a small low coral island. The obsidians are from excavations of a Lapita pottery complex site (Egloff, 1975: 15), which elsewhere have been radiocarbon dated to around 3000 years ago. This site also contained obsidians from the Talasea source which will not be considered here. The distribution pattern (Figure 1.5) likens the site to the mid levels of the Kohin Cave site. Lesu is a New Ireland coastal village site with several refuse mounds (White, 1972: 309). Lou Island source obsidian from one of these, mound V, was analysed and is presented in Figure 1.5. The carbon date for the base of mound V is 2460±250 (ibid.). The Lous obsidian source pattern from layers 1-3 at Lesu is similar to the middle and lower levels of Umleang (Figure 1.2). The lowest level 4 is similar to the mid zones of the Emsin site layer 4.

Ahus Island

| Layer | 1 | 2 | 3 | 4 | 5 | 6 | 7 | 8 | 9 Groups |
|---|---|---|---|---|---|---|---|---|---|
| 1 | | | | | | ▲▲▲ | ▲▲▲▲<br>▲▲ | | |
| 2 | ▲▲▲▲<br>▲ | ▲ | | | | | ▲ | | |
| 3 | ▲▲▲▲<br>▲▲ | ▲▲ | | ▲ | | | | | ▲ |
| 4 | ▲▲▲▲<br>▲▲▲ | | | ▲ | ▲▲ | | | | |
| 5 | ▲▲▲▲<br>▲▲▲ | ▲▲ | | | | | | | |
| 6 | ▲▲▲▲<br>▲▲▲▲<br>▲▲ | ▲ | | ▲ | | | | | |
| 7 | ▲▲▲▲<br>▲▲▲▲<br>▲ | ▲ | | | | | | | |
| 8 | ▲▲▲▲<br>▲▲ | | | | | | | | |
| 9 | ▲▲▲▲<br>▲▲▲▲ | ▲ | | ▲ | | | | | |
| 10 | ▲▲▲▲<br>▲ | | | ▲▲▲▲ | | ▲ | | | |
| 11 | ▲▲▲▲<br>▲ | | | ▲ | | ▲▲ | | | ▲ |

Figure 1.3 The sources of obsidian artefacts from the archaeological site on Ahus Island

Kohin Cave

| Layer | 1 | 2 | 3 | 4 | 5 | 6 | 7 | 8 | 9 Groups |
|---|---|---|---|---|---|---|---|---|---|
| 1 | | ▲ | | ▲ | | | ▲▲ | | ▲ |
| 2 | ▲ | | △ | ▲ | | ▲ | ▲▲▲▲ | | |
| 3 | ▲ | ▲ | ▲ | ▲▲▲ | ▲ | ▲▲ | ▲▲▲▲<br>▲▲▲△<br>△ | △ | ▲ |
| 4 | △△△△<br>△△ | △△ | ▲▲▲ | ▲▲ | | ▲▲▲▲<br>▲▲▲▲<br>▲▲▲▲<br>▲▲▲△△ | ▲▲▲▲<br>▲▲▲▲<br>△ | | |
| 5 | ▲ | | | | | | ▲ | | |
| 6F | | ▲▲ | ▲▲▲▲<br>▲▲▲ | | | ▲ | ▲△ | | |
| 6 | | ▲▲ | ▲△ | | | ▲▲▲ | ▲▲▲ | | |
| 8 | | ▲△△ | ▲▲▲▲<br>▲▲ | | | ▲▲▲ | ▲▲▲▲<br>▲▲▲▲<br>△△△△ | | ▲△ |
| 9 | ▲ | | ▲▲▲▲<br>△ | | | | ▲▲△ | | |

Figure 1.4 The sources of obsidian artefacts from Kohin Cave

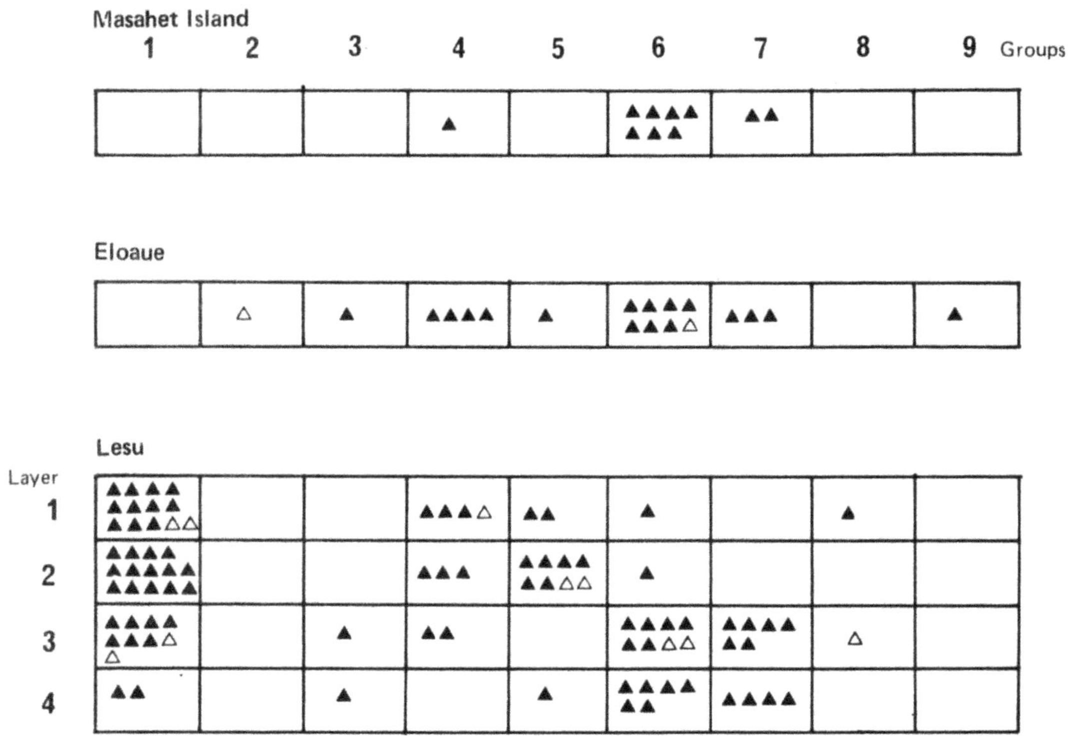


Figure 1.5 The sources of obsidian artefacts from Masahet Island, Eloaue, and Lesu

The brief discussion of the preliminary results from a detailed analysis of Lou-Pam obsidians shows some encouraging consistencies in the changing patterns of source exploitation. The full discussion of the results from this study must await the further analysis of another 1000 specimens from surface and excavated sites in the region.

The analytical technique of PIXE-PIGME applied to the non-destructive chemical analysis of obsidians, and the use of multivariate computer treatment of the data, have now made it possible to consider more detailed questions of communication and interaction in a prehistoric Melanesian coastal situation. Further work is proceeding along these lines.

ACKNOWLEDGEMENTS

We are very grateful to the following for the assistance with this project: The Australian Institute of Nuclear Science and Engineering for supporting the research with grants. The staff at the Lucas Heights 3 meV accelerator provided invaluable support. J. Kennedy, J. P. White and B. J. Egloff, provided excavated specimens and discussed various results, and Y. Pittelkow assisted with computer programmes and statistical advice.

REFERENCES

Ambrose, W. R., 1976a. Obsidian and its prehistoric distribution in Melanesia. Pp. 351-378 In: N. Barnard (Ed.), Ancient Chinese bronzes and Southeast Asian metal and other archaeological artifacts. National Gallery, Melbourne.

Ambrose, W. R., 1976b. Intrinsic hydration rate dating of obsidian. Pp. 81-105 In: R. E. Taylor (Ed.), Advances in obsidian glass studies, archaeological and geochemical perspectives. Noyes, New Jersey.

Ambrose, W. R., 1978. The loneliness of the long distance trader in Melanesia. Mankind 11(3): 326-33.

Bird, J. R., Russell, L. H., Scott, M. D. and Ambrose, W. R., 1978. Obsidian characterization with elemental analysis by proton induced gamma-ray emission. Analytical Chemistry 50(14): 2082-2084.

Bird, J. R. and Russell. L. H., 1976. Applications of prompt nuclear analysis techniques to the study of artefacts including southwest Pacific obsidian. Pp. 317-336 In: N. Barnard (Ed.), The proceedings of a symposium on scientific methods of research in the study of ancient Chinese bronzes and Southeast Asian metal and other artifacts. National Gallery of Victoria, Melbourne.

Blake, D. H. and Ewart, A., 1974. Petrography and geochemistry of the Cape Hoskins volcanoes, New Britain, Papua New Guinea. Geological Society of Australia Journal 21: 319-31.

Blake, D. H., 1976. Pumiceous pyroclastic deposits of Witori Volcano, New Britain, Papua New Guinea. In: R. W. Johnson (Ed.), Volcanism in Australia. Elsevier, Amsterdam.

Blake, D. H. and McDougall, I., 1973. Ages of the Cape Hoskins volcanoes, New Britain, Papua New Guinea. Journal of the Geological Society of Australia 20(20): 199-204.

Doran, J. E. and Hodson, F. R., 1975. Mathematics and computers in archaeology. Edinburgh University Press.

Duerden, P., Cohen, D. D., Clayton, E., Bird, J. R., Ambrose, W. R. and Leach, B. F., 1979. Elemental analysis of Thick Obsidian Samples by Proton Induced X-ray Emission Spectrometry. Analytical Chemistry 51: 2351-4.

Duerden, P., Bird, J. R., Scott, M. D., Clayton, E., Russell, L. H. and Cohen, D. D., 1980. PIXE-PIGME studies of artefacts. Nuclear Instruments and Methods 168: 447-452.

Egloff, B. J., 1975. Archaeological investigations in the coastal Madang area and on Eloaue Island of the St Matthias Group. Records of the Papua New Guinea Public Museum and Art Gallery 5.

Johnson, R. W., Smith, I. E. M. and Taylor, S. R., 1978. Hot-spot volcanism in St Andrew Strait, Papua New Guinea: geochemistry of a Quaternary bimodal rock suite. BMR Journal of Australian Geology and Geophysics 3(1): 55-69.

Johnson, R. W. and Smith, I. E. M., 1974. Volcanoes and rocks of St Andrew Strait, Papua New Guinea. Journal of the Geological Society of Australia 21: 333-51.

Johnson, R. W. and Blake, D. H., 1972. The Cape Hoskins area, Southern Willaumez Peninsula, the Witu Islands, and associated volcanic centres, New Britain: volcanic geology and petrology. Bureau of Mineral Resources, Geology and Geophysics. Record 1972/133.

Kennedy, J., 1980 (in press). Prehistoric research in Manus Province. Research in Melanesia.

Klecka, W. R., 1975. Discriminant analysis. Pp. 434-467 In: Nie et al., 1975, q.v.

Lowder, G. G. and Carmichael, I. S. E., 1970. The volcanoes and calders of Talasea, New Britain: geology and petrology. Geological Society of American Bulletin 81: 17-38.

Mallick, D. I. J., 1973. Some petrological and structural variations in the New Hebrides. In: P. J. Coleman (Ed.), The Western Pacific, island arcs, marginal seas, geochemistry. University of Western Australia Press.

Neilson, K. K., Hill, M. W., Mangelson, N. F. and Nelson, F. N., 1976. Elemental analysis of obsidian artifacts by proton particle-induced X-ray emission. Analytical Chemistry 48(13): 1947-1950.

Nie, N. H., Hull, C. H., Jenkins, J. G., Steinbrenner, K. and Brent, D. H., 1975. Statistical package for the social sciences. 2nd edition.

Ottaway, B., 1974. Cluster analysis of impurity patterns in Armorico-British daggers. Archaeometry 16(2): 221-231.

Parkinson, R., 1907. Dreissig Jahre in der Südsee. Strecker and Schröder, Stuttgart.

Pain, C. F., 1980. Stratigraphy and chronology of volcanic ash beds on Lou Island, Papua New Guinea. (In press).

Renfrew, C., 1969. Trade and culture process in European prehistory. Current Anthropology 10: 151-169.

Reynolds, M. A. and Best, J. G., 1976. Summary of the 1953-57 eruption of Tuluman volcano, Papua New Guinea. In: R. W. Johnson (Ed.), Volcanism in Australia. Elsevier, Amsterdam.

Smith, I. E., 1973. Late Cainozoic Volcanism in the Southeast Papuan Islands. Bureau of Mineral Resources, Geology and Geophysics. Record 1973/67.

Smith, I. E., 1974. Obsidian sources in Papua-New Guinea. Archaeology and Physical Anthropology in Oceania 9(1): 18-25.

Specht, J. R., 1974. Lapita pottery at Talasea, West New Britain, Papua New Guinea. Antiquity 48: 302-306.

Wall, T., 1976. Use of the Research Reactor MOATA and associated facilities for the source identification of obsidian artifacts. Pp. 337-350 In: N. Barnard (Ed.), Ancient Chinese bronzes and Southeast Asian metal and other archaeological artifacts. National Gallery, Melbourne.

White, J. P., 1972. Carbon dates from New Ireland. Mankind 8(4): 309-310.

White, J. P., Downie, J. E. and Ambrose, W. R., 1978. Mid-Recent human occupation and resource exploitation in the Bismarck Archipelago. Science 199: 877-879.

White, J. P., and Modjeska, N., 1978. Acquirers, users, finders, losers: the use axe blades make of the Duna. Mankind 11(3): 276-287.

Wishart, D., 1971. IA Manual. Clustan User. University of St Andrews, Scotland.

## 2. A NEW DEVELOPMENT IN UNDERSTANDING THE PREHISTORIC USAGE OF NEPHRITE IN NEW ZEALAND

Russell J. Beck

### INTRODUCTION

Nephrite, although not a particularly hard stone is, because of its felted structure, extremely tough with almost metal qualities, and its discovery presented the prehistoric Maori of New Zealand with a revolutionary new material for tool making.

Early examples of the use of nephrite and semi-nephrite show the percussion method of manufacture, but because of the toughness of the material the task would have been an arduous one, and this technique, so well established for comparatively brittle stone types such as basalt, argillite etc., no longer applied. As a consequence, a new technology which involved sawing the stone developed. The superior properties of nephrite were quickly recognized and usage became widespread throughout the country with far reaching cultural effects. The aesthetic qualities of nephrite did not go unnoticed and some particularly fine art objects such as ceremonial adzes, weapons and ornaments were produced. With the introduction of metal from contact with the European in the protohistoric period, the production of nephrite tools ceased, and the manufacture of ornaments for trading, notably hei-tiki, flourished. Desire to possess metal caused the sacrifice of many neprhite adzes which were re-worked into tiki and to some degree the adze shape dictated the development of the familiar hei-tiki design (Skinner, 1974). All too soon the knowledge of many source locations of nephrite, and much of the technology employed in working it, was not recorded and therefore lost.

The New Zealand sources of nephrite and semi-nephrite all lie in the South Island, and can be grouped into five areas: Nelson, Westland, South Westland, Otago and Southland. Generally, the Westland source affords the hard dark green nephrite in abundance, and the other locations are better known for the paler colours of predominantly semi-nephrites (Beck, 1970). It appears that the Maori exploited all of these sources (except those in Southland), and possibly others yet to be located. The major Otago source was re-discovered as late as 1970, and now has full protection as an archaeological site. This is the only known undisturbed prehistoric source (Beck, 1977).

Nephrite, otherwise known as Jade, is a compact felted variety of the tremolite actinolite series of amphiboles, being essentially a calcium magnesium silicate composition. The unique toughness of nephrite results from the structure, which is composed of felted, entangled twisted crystal fibres characteristically known as nephritic. Semi-nephrite differs from nephrite in that the felted structure is not completely developed throughout, and many crystal fibres instead of being entwined are coarsely crystalized, or orientated in a parallel arrangement, giving it a schistose appearance (Turner, 1935). In New Zealand, the colour of nephrite is generally a rich grass green, but can grade from very dark green to white, depending on the ferrous iron content. Oxidized varieties are yellow, orange or brown, and owe their colour to ferric iron.

An on-going project in my study of New Zealand nephrite has been to trace the sources of the nephrite used by the prehistoric inhabitants. Some success has been achieved in establishing the probable sources, plus the discovery of several new occurrences that appear not to have been known or utilized by the Maori.

Although some nephrite has distinguishing characteristics that allow reasonably accurate sourcing, many locations produce a whole variety of material which can overlap in appearance with other sources. Complicating the sourcing task further, is the presence of a group of artefacts which have changed their physical appearance through natural weathering or as the result of suspected heating in a fire. This group of artefacts can be distinguished by their colour, which is usually brown, or yellowy shades, or chalky greyish-white.

FIRING TESTS

In an attempt to duplicate the appearance of these artefacts, a selection of samples of nephrite and semi-nephrite from known sources were placed in an open fire and removed when reaching red heat, then left to cool in the atmosphere. The heated samples presented an outside appearance of black to brownish; however, when this dark skin was removed by grinding the inside colour was seen to have changed dramatically.

A few samples had turned completely brown, while others went silvery-white, resembling unheated specimens from different sources. Furthermore, several heated specimens from a known source now resembled some prehistoric artefacts previously believed to be natural, and from an unknown source. More surprising was the fact that during the grinding off of the black coating after firing, the softer semi-nephrite specimens appeared to have increased their hardness (Beck, 1978).

It was clear that these preliminary results had important implications and this gave encouragement for further experiments. Over a period of several months, a large number of test pieces were fired, using both the open fire and an electric kiln. Samples were prepared by selecting typical specimens from eight known sources of predominantly semi-nephrite from which 8 mm-thick slabs were sawn. In order to present the softest face for treatment and testing, the slabs were cut parallel to the schistosity plane of the stone. These were then cut into several matching pieces approximately 20 mm x 20 mm from each locality. All pieces were then ground flat and smooth with a diamond lap and engraved with catalogue numbers together with face marks to assist matching. One sample of each pair was then fired.

Samples were exposed to heat for varying periods of time up to two days, and temperatures up to 1000°C. The higher temperatures destroyed the structure of the stone causing crazing, and the specimen became opaque. The low fired specimens (300°C) showed a small change, mainly in colour.

By a process of elimination it was found that 650° C for a short period of time was the most effective temperature which did not damage the specimen too much, and yet produced an appreciable change in colour and increase in hardness. Another reason for choosing this temperature was that with the resources available to the Maori, a temperature of 650° C was easily attainable using a camp fire of native timbers.

Further experiments were carried out solely in an electric kiln and the following examples describe the changes in appearance.

W301 A sample of high quality Westland nephrite was heated to 1000° C and the surface ground off. The colour changed to an opaque greyish-white with the structure destroyed and crazed, with small cracks outlined by a dark rusty colour.

W302 A somewhat oxidized sample of Westland nephrite grading from yellow to green was heated to 650° C and ground smooth. The colour changed to a reddish brown, grading darker towards the end coinciding with the green of the unheated portion.

W106 A softer olive green semi-nephrite from Westland was heated to 650° C, producing the typical dark outer skin. A further sample from the same piece was heated to 1000° C, giving the typical characteristics for this temperature: outer rusty colouring, opaque grey inside, and the structure destroyed with crazing (See Figure 2.1).

The following samples were all heated to only 650° C over a period of an hour and left to cool naturally with the surface ground off to reveal the inner colour.

W103 A Westland sample (displaying a natural olive coloured oxidized rind) became brown with the inner portion turning silvery whitish-green. Under the microscope the light silvery colour is produced by light reflecting off a multitude of small uneven feather fractures throughout the sample, parallel to the plane of schistosity. The brown is probably due to the further oxidization of the outer area, which was naturally more oxidized than the centre. This feature was quite constant throughout all the experiments, and one could surmise that any olive green nephrite would change to dark brown on heating. The minute feather fractures that cause the white portion are also present in the brown area, but masked to some degree by the dark colour (See Figure 2.2). Similarly, an Otago sample (D2) with the natural colouring of pale green and olive green, upon firing, became silvery greenish-white, with broad dark brown areas. Its appearance closely matched a prehistoric artefact (D46. 1053), previously believed to be from an unknown source because of its unusual, assumed "natural" colouring (See Figure 2.3). One can therefore postulate what the original stone colouring was of this obviously heated artefact by applying the general rule that dark brown areas would have originally been olive green, and the light portion pale green. It was interesting to note that the particular specimens prepared for the Vickers hardness tests were polished prior to firing, and this tended to inhibit the formation of the customary dark skin on the surface.

Many heated samples showed a veining effect where the structure opened, forming tiny cracks which became conspicuous by the presence of a dark coloured residue in them. Correspondingly, many prehistoric artefacts show this characteristic black veining effect, which provides another distinguishable feature to establish whether an artefact had been heated (See Figure 2.4).

Figure 2.1:W106, Westland semi-nephrite.Left-natural specimen,olive green.Centre portion heated to 650°C showing black outer skin in contrast to lower portion where the surface has been removed.Right portion heated to 1000°C with surface ground off to reveal a crazed opaque grey interior with a rusty exterior. Scale full size.

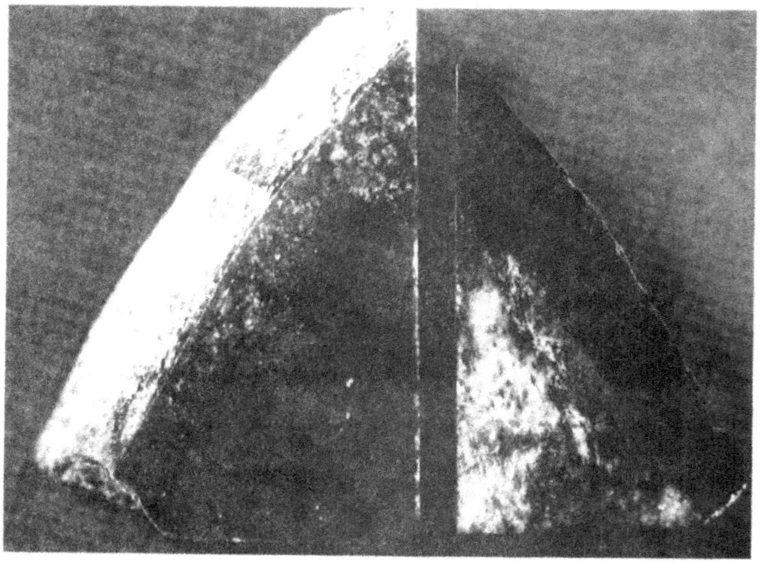

Figure 2.2:W103,Westland semi-nephrite.Left-natural specimen, medium green and olive green rind.Right,heated portion, showing white area which was previously green and olive shade, now dark brown. Full size.

Figure 2.3:Left(D2)-natural Otago semi-nephrite, olive and pale green. Centre-heated portion of D2 showing dark brown and light areas. Right, prehistoric artefact(D46.1053) with similar markings and colour to the heated portion.This artefact was previously thought to be of a natural colouring.Scale, full size.

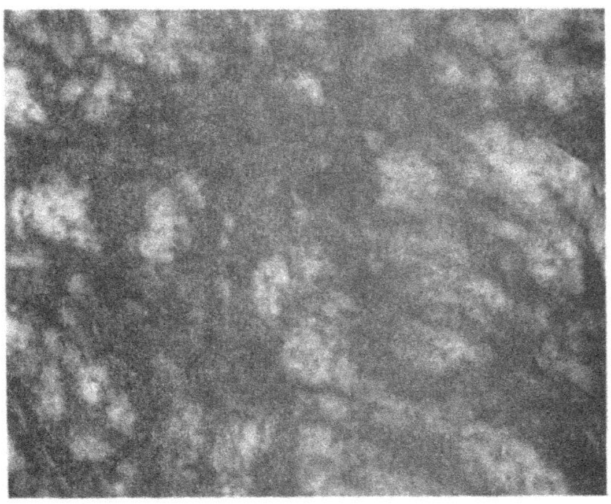

Figure 2.4:W300.A heated Westland sample showing black veining effect which is often present in many artefacts suspected of being heated. Scale, twice size.

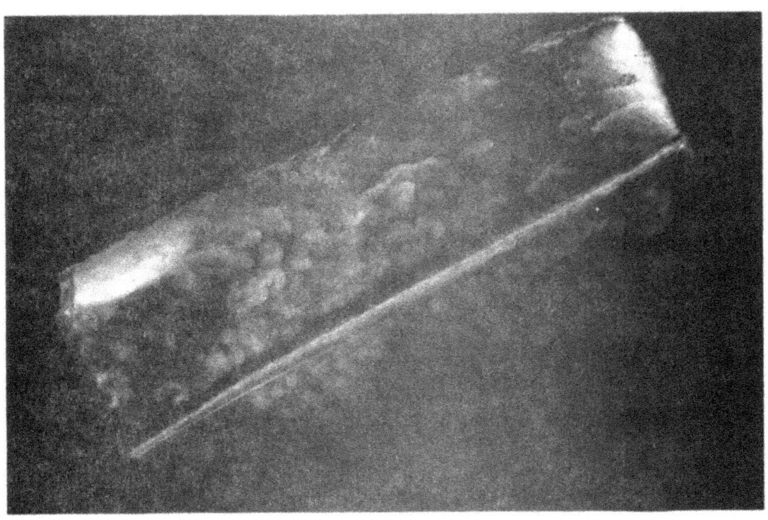

Figure 2.5:A prehistoric Maori nephrite artefact(D46.1300) exhibiting circular feather features which were probably induced by localised heating or quenching. Scale, full size.

A further diagnostic characteristic is a series of concentrations of large circular feather fractures (See Figure 2.5), and experiments on samples from several sources revealed that this effect could be simulated by uneven localised heat or partial quenching in cold water. Nephrite often contains naturally produced feather fractures, but generally does not have this circular fish scale effect.

HARDNESS

Hardness is a variable property with most minerals and its measurement can pose difficulties. The Moh scale, an arbitary scale of 1-10, is most commonly used and is gauged by the ability of a harder mineral to scratch a softer one. Compact nephrite when tested on a smooth surface is $6\frac{1}{2}$ on the scale; however, there are considerable variations depending on structure and direction. Varieties displaying a definite grain, in particular schistose semi-nephrites, give least hardness values on the surface parallel to the schistosity, and greatest on the end grain. A hardness of $6\frac{1}{2}$ on the end grain and as low as 5 on the schistosity face is not uncommon for semi-nephrite.

Testing of the heated and unheated samples was done as previously mentioned on the softest face, and initially by the scratch method. Using the matched pairs of samples it was found that as a rule, the heated specimen would scratch the unheated one, but the natural one would not mark the heated sample. Care was taken to check these scratch marks under a microscope to make sure that the white scratch mark was in fact a furrow and not a deposit of the softer material adhering to the surface of a harder sample. The hardness tests were carried out by scratching the samples, using hardness testers of known hardness, and an impact apparatus and the Vickers Micro Indentation Hardness Test. This latter instrument produces a minute square indentation on the surface of the specimen with a diamond. The indentation is then measured with a micrometer microscope, and scaled to represent kgs per $mm^2$ (McCawley, 1978).

For the Vickers test, nephrites and semi-nephrites from many sources were used, and prepared as before but of a smaller size. As semi-nephrite can contain inclusions of accessory minerals and variations in compactness, homogeneous areas no larger than 20 mm square were selected and sawn in two giving matched consistent pairs. The surfaces were polished to allow as accurate a measurement as possible, and one of each pair was then fired. Because of the multi-crystalline structure of the samples, four readings on different parts of each specimen were taken and averaged. Difficulty was experienced with some samples which did not always give measurable readings due to fracturing on the surface, and several attempts were necessary. Conversely, the fired samples were consistent with sharper indentations. All fired samples showed an increase in hardness over their matched unheated counterparts, varying from small to larger amounts but averaging an increase of 50% over 20 samples with 80 readings. On the whole, semi-nephrite gave a greater increase, and in some instances there was an indication of a case hardening effect. Evidence suggests that the heated specimens lose some of their toughness or flexibility, with a tendency to become more brittle. This may have some bearing on the apparent increase in hardness, and further work is in progress to establish what structural and mineralogical changes are taking place. Preliminary results from X-ray diffraction analysis of powdered samples show little, if any, change with the samples heated to 650° C; however, the samples taken to 1000° C have changed mineralogically from tremolite to close to diopside.

Another interesting effect observed was a weight loss. Samples from four sources were carefully weighed before and after firing, and all revealed a weight loss averaging 0.34%, with semi-nephrites giving a greater weight loss than compact nephrite. This may be due to the removal of some of the water which nephrite contains.

It was also noted that specimens of semi-nephrite which presented difficulties in polishing were, upon firing, able to be polished effortlessly to a much higher degree.

A well known property of nephrite is its ability to "ring". Suitably shaped pieces of suspended nephrite will, when struck, hold a tone like a gong for several seconds. Semi-nephrites seldom have this property, but after firing, specimens lost their dull sound and assumed a decided "tinkle" when handled

IMPLICATIONS

There is no doubt that a large number of prehistoric nephrite artefacts were fired. It is certain that the Maori was aware of the resulting effects and took advantage of them.

Two major results are evident when nephrite is subjected to heat: a change in appearance, and an apparent increase in hardness. Let us surmise what the implications could be.

There are strong indications that a predominant reason for heating nephrite was to facilitate the colour change; yet one may ask—why transform an attractive translucent green object to a milky white colour? The Maori, like people of other cultures, deeply appreciated the aesthetic qualities of nephrite, and the various types and colours were named and prized for their beauty and rarity. Although New Zealand is world famous for the bright green shade, the Maori held the rarer pale green silvery variety (inanga) in high reverence; so much so that long journeys to the interior of the South Island were undertaken to procure this variety alone (Heaphy, 1959). Inanga received its name from its similarity in colour to a juvenile river fish (Galaxias attenuatus), and although not as colourful, has a beauty and character of its own which is seldom equalled by the green varities. It is interesting that a fish theme was an important part of Maori mythology on nephrite. This feature, together with the beauty and relative rarity of inanga, make it hardly surprising that artefacts of this material were held in great esteem.

With suitable specimens we have seen that the firing technique can duplicate a material close to the inanga appearance, and an opportunity to do this would have been seized upon by the prehistoric Maori. The increase in hardness provided a dual benefit. Undoubtedly, many artefacts suffered heating accidentally, and a few show overheating to the point of destruction. Some indicate signs of slight heating and possibly valued pieces were fired carefully in stages to obtain the desired colour, as there was no guarantee that the anticipated effect would be achieved. In addition, other colours, such as an attractive yellow/orange, and a variety with reddish markings, were possible and appreciated. It would appear that other minerals were also heated to effect a colour change; for example brown limonite was altered to red hematite and used as a paint pigment (Beck, 1975).

Early communications with the European in the protohistoric period regarding the source of nephrite record: "When first dug from its bed it was found to be of a soft nature but hardened on exposure to the air" (Polack, 1838); and another reference adds: "[It] must be wrought in its soft state" (Wilkes, 1845). Likewise, a legend which implies nephrite was to be found in a large fish, states that when first taken it was soft and became hard on exposure (Wilkes, 1945). These may or may not relate to the heating process but essentially they do illustrate that the prehistoric Maori was aware of a hardening effect for nephrite. There appear to be obvious advantages if a softer semi-nehprite could be worked relatively easily, then heated and brought up to a hardness equalling a compact nephrite. The large number of predominantly semi-nephrite artefacts which display fired characteristics offer evidence that this technique was practised.

On the other hand, neprhite tools were commonly fashioned with the grain longitudinally, giving the hardest surface at the cutting edge, and an increase in hardness from firing, possibly accompanied by a small loss in toughness, may not have given as great a practical advantage as expected.

In conclusion, whether nephrites were heated by the prehistoric Maori accidentally, or deliberately for whatever reason, the experiments described here have revealed additional information for assisting in the sourcing of artefacts, and interesting aspects which may be applicable in other instances and directions.

ACKNOWLEDGEMENTS

My thanks to Dr. M. Duggan, Geology Department, University of Otago, for assistance and use of hardness testing equipment and X-ray analysis. I also thank the Management and laboratory staff of New Zealand Aluminium Smelters Ltd for the use of their hardness testing facilities.

I am grateful to Dr. A. P. Poole, Chairman of the Southland Museum Trust Board (Inc), for his encouragement; also to Dr. B. F. Leach, Anthropology Department, University of Otago, for his valuable comments, assistance with apparatus and the reading of this paper at Khabarovsk on my behalf.

REFERENCES

Beck, R. J., 1970. New Zealand Jade, the story of greenstone. A. H. and A. W. Reed, Wellington.

Beck, R. J., 1975. Probable sources of Southland stone artefacts. Unpublished paper presented to the Otago Anthropological Society's 1975 Conference.

Beck, R. J., 1977. Slip Stream nephrite Dart Valley Otago. AGMANZ News 8(1): 6-8.

Beck, R. J., 1978. A new development in Maori nephrite technology. Unpublished paper, Southland Museum.

Heaphy, C., 1959. Notes of an expedition to Kawatiri and Araura, on the western coast of the Middle Island. Pp. 203-248 In: N. M. Taylor (Ed.), Early Travellers in New Zealand. Clarendon Press, Oxford.

McCawley, E. L., 1978. How, where and why hardness varies in gemstones. Lapidary Journal 32(5):1090-1105.

Polack, J. S., 1838. New Zealand. Volume 1. Bentley.

Skinner, H. D., 1974. Maori amulets. Pp. 45-98 In: Gathercole, P., Leach, B. F. and H. M. Leach (Eds.), Comparatively speaking. University of Otago Press, Dunedin.

Turner, F. J., 1935. Geological investigations of the nephrites, serpentines and related greenstones used by the Maoris of Otago and South Canterbury. Transactions of the Royal Society of New Zealand. 65: 187-210.

Wilkes, C., 1845. Narrative of the United States Exploring Expedition. Volume 2. Lea and Blanchard.

# 3. PACIFIC OBSIDIAN CATALOGUE

J. R. Bird, P. Duerden, W. R. Ambrose and B. F. Leach

## INTRODUCTION

Obsidian is ideally suited to provenance studies using elemental analysis and numerous techniques have been applied for this purpose. It is not surprising, therefore, that as a new analytical technique becomes available it will usually find useful application for obsidian characterization. Such studies include the search for similarities among groups of artefacts and comparisons with known source material, so that plausible suggestions can be made as to the geographical origin of artefact groups.

Measurement of a small number of elements, and even of such physical parameters as density and refractive index, may suffice to demonstrate differences between a small number of samples. Nevertheless, similarities must be shown to apply over all available parameters and it is difficult to ensure in advance that any programme of measurements will provide adequate information for the problem in hand. There are thus a number of reasons for cataloguing the composition of source materials and separable artefact groups:

(a) to facilitate the planning of measurements to tackle specific problems;
(b) to allow the probability of ambiguities arising to be assessed and minimized;
(c) to encourage standization of measuring techniques;
(d) to allow regional studies to be placed in a context of a broader understanding of trends in obsidian composition; and
(e) to provide links with studies of the composition of related volcanic materials.

### Analytical Techniques

Neutron activation analysis (NAA) has a well-established role in the achievement of high sensitivities for measuring a variety of elements, particularly of rare earth elements. It has been used for numerous obsidian studies and, in the case of Pacific obsidian, results have been reported by Wall (1976) for source materials from the Papua New Guinea region, and by Leach and Warren (1981, and private communication) for New Zealand and Oceanic obsidian. While NAA can be, and has been, applied to large numbers (thousands) of samples, samples are restricted to a size and shape suitable for irradiation in a reactor, and repeated measurements may be necessary following irradiation.

Atomic absorption spectroscopy (AAS) can be readily applied in the laboratory, but it is a destructive technique which can be time-consuming if many elements are to be analysed. AAS has been used for the analysis of Pacific obsidian by Reeves and Ward (1976).

X-ray fluorescence (XRF) is also used in many laboratories to provide a multi-element capability which can be easily applied to the study of artefacts. The number of observable elements is limited only by the detector efficiency, the energy of exciting radiation, and the presence of X-rays from major elements which may limit the sensitivity for other neighbouring X-ray energies. The precision of XRF can be affected by sample shape and position. XRF has been used in a number of important studies of the distribution of Pacific obsidian by Smith, Ward and Ambrose (1977), and of related volcanic rocks by Johnson, Smith and Taylor (1978). The development of a portable system for use in the field or museum promises to be of considerable importance to such studies (McCallum et al., 1979; and private communication).

Proton induced gamma-rays (PIGME) have been found to be useful for rapid, non-destructive analysis of obsidian (Bird and Russell, 1976), although only three elements are normally determined, viz., fluorine, sodium and aluminium. This is a precise technique which can be applied to almost any sample shape and it is not seriously limited by artefact weathering. However, it does require an ion accelerator which is usually available only in specialized laboratories. Measurements on all known obsidian sources from Western Melanesia and approximately 700 artefacts from that area have confirmed the value of this technique (Bird et al., 1978).

Proton induced X-rays (PIXE) can also be used for rapid, non-destructive multi-element analysis, of either specially prepared digests of obsidian (Nielson et al., 1976) or normal samples, including artefacts (Duerden et al., 1979). The spread in results obtained from one type of obsidian, arising from variations in surface quality, X-ray attenuation, etc., is more than offset by the fact that at least 20 elements can be determined, some of which are likely to show major changes in concentration from one type of obsidian to another.

## PIXE-PIGME METHOD

Following the successful application of these two techniques, equipment has been developed for their use simultaneously for studying major, minor and trace elements in obsidian (Duerden et al., 1980). Measurements have been carried out on all known sources of obsidian from the South Pacific and neighbouring areas and, in this paper, progress is reported on the analysis of these data and comparison with results of other work. The aim of the project is to provide a catalogue of compositional information which will serve as a reference for future studies of obsidian artefacts. It is also hoped to compare measurements using NAA and XRF techniques on the same samples, so that a choice of analytical methods will be available in the future, whose performance and results are related through the catalogue.

### Measurements

Details of the technique have been reported elsewhere (Duerden et al., 1980) and the following summary relates specifically to the obsidian catalogue work. A beam of 100 to 200 nA of 2.5 MeV protons is used to irradiate each obsidian sample in turn and gamma-ray and X-ray detectors measure the radiation emitted during irradiation (see Figure 3.1). A typical measuring time is 5 minutes per sample and automatic sequencing permits up to two or three hundred samples to be studied per day with only limited effort necessary for periodic sample loading. However, quite a few hours are needed to set up the equipment and check all aspects of its operation so that the method is best applied for the intercomparison of large sets of samples.

Samples are mounted on metal plates (See lower half of Figure 3.1) before the experiment, and a special carrier holding up to 60 plates is inserted into the accelerator vacuum system for each measuring cycle. Each carrier is washed in distilled water or nanograde benzene using ultrasonic agitation, but no other sample preparation steps are necessary. The maximum sample size which can be loaded in this way is such that it will pass through a 5 cm diameter loading tube. Larger samples can be mounted, one at a time, within the main vacuum chamber which is 20 cm diameter.

<u>Typical Results</u>

A typical gamma-ray spectrum is shown in Figure 3.2 from which it can be seen that peaks from Na and Al gamma-rays, as well as a broad spectrum region from F gamma-rays are the dominant features. Other elements such as Li or B can be detected with longer running times.

A typical X-ray spectrum is shown in Figure 3.3, taken with a special pinhole filter to allow simultaneous observation of low energy X-rays from the major elements (from Na to Ca), as well as higher energy X-rays from minor and trace elements (from Ti to Nb). The elements normally determined from this spectrum include Al, Si, Cl, K, Ca, Ti, Cr, Mn, Fe, Cu, Zn, Rb, Sr, Y, Zr and Pb. However, additional elements can be determined if they are present in significant quantities.

<u>Anomalous Results</u>

When studying artefacts, a few per cent of the samples may give anomalous results, even although measurements are made on an area of the surface which appears free from defects. Such samples are remounted with additional care to find a more suitable part of the surface and, in general, most of these then give normal and reproducible results. However, up to 1% of artefacts may give such unreliable results that little useful information can be obtained from them. Such samples would require preparation of a freshly exposed surface, which can be as small as 1 mm in diameter, or even less, or the use of other analytical techniques.

<u>Data Processing</u>

Information can usually be obtained from the two spectra on at least 15 elements, either by simple data reduction during the experiment or by more sophisticated processing after the experiment is complete. For the latter purpose, the complete spectra can be saved on a computer disk file, while the numerical data from on-line processing can also be automatically transferred to a special archive file. The results of on-line processing can also be entered into the archive file which provides facilities for sorting data and runs and preparing listings and plots of selected information. It can also be used to carry out arithmetic operations, such as ratio calculations, normalization, etc., as well as to obtain distributions and statistical parameters for selected sets.

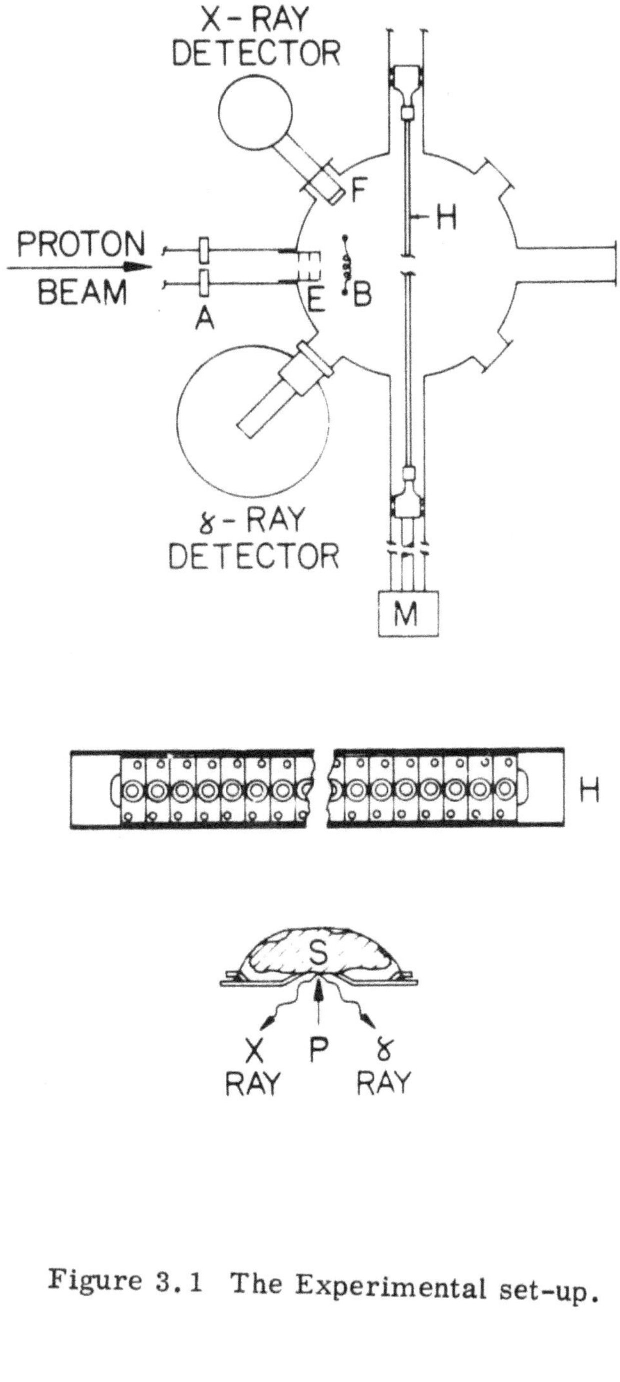


Figure 3.1 The Experimental set-up.

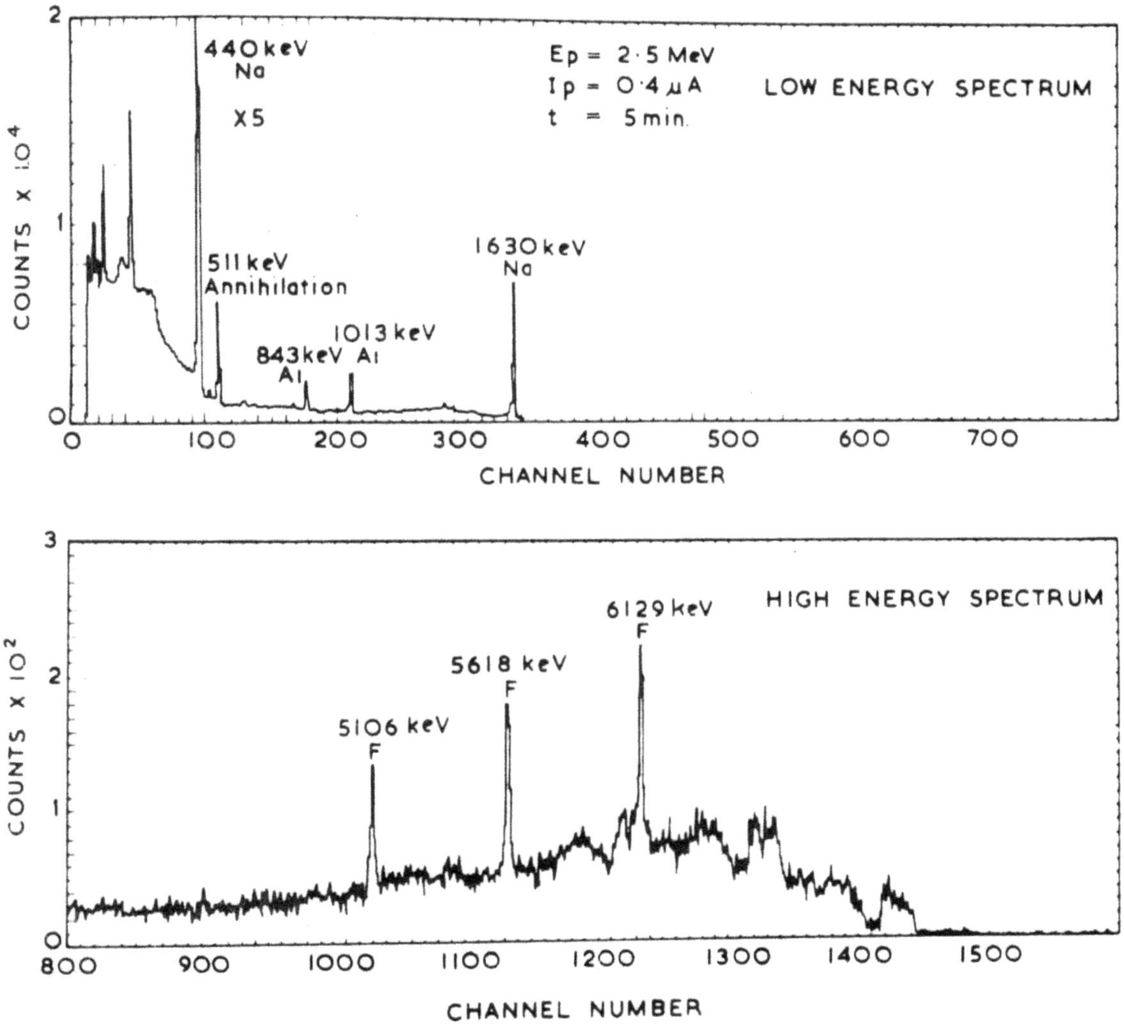


Figure 3.2 Typical gamma-ray spectrum from obsidian.

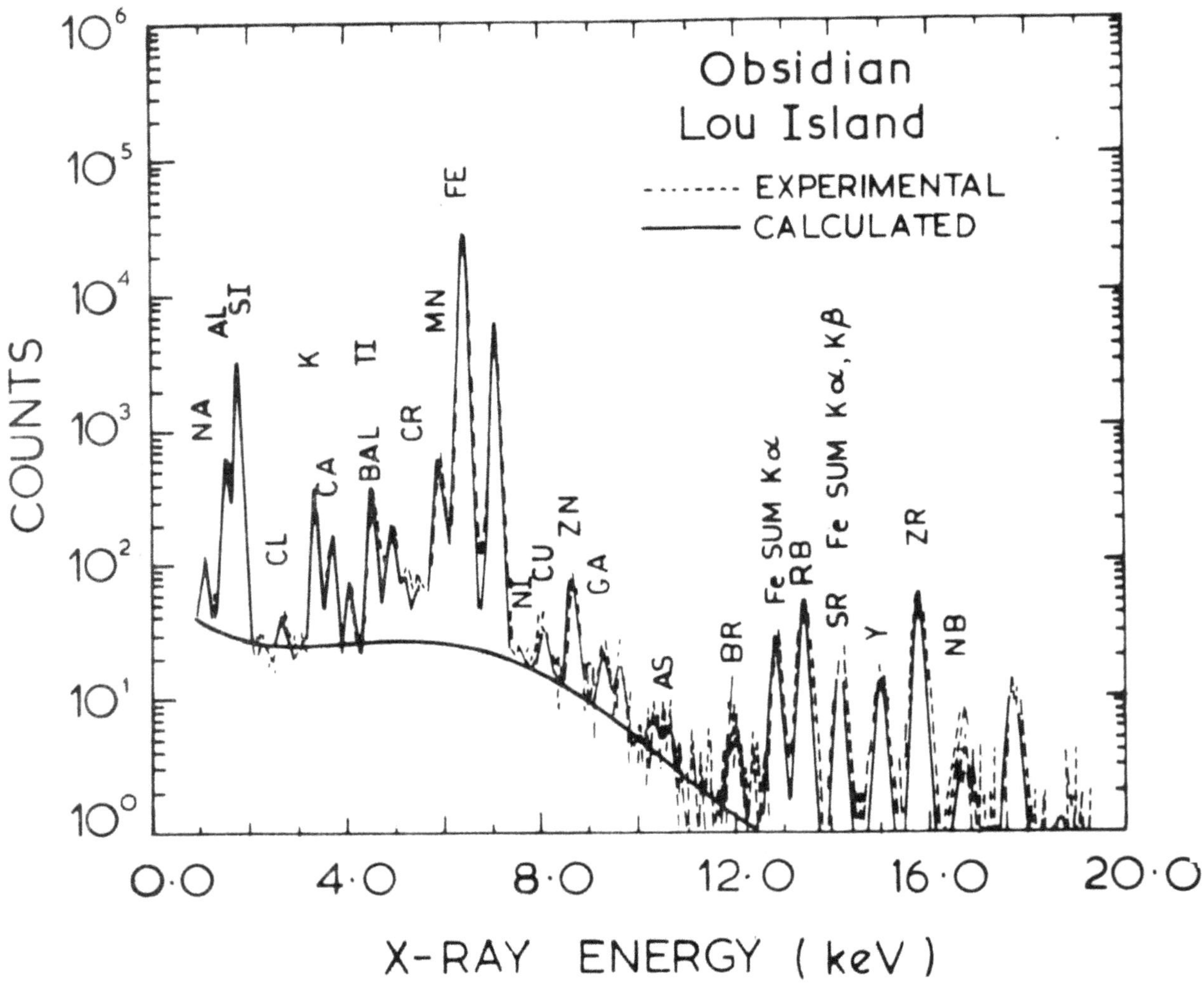


Figure 3.3 Typical PIXE spectrum.

In order to obtain absolute values of element concentrations, standard samples are included in a sequence of measurements, and the results used to calculate normalized values after processing to correct for background and multiple peaks. However, this may not always be necessary, and greater accuracy can be obtained for relative values between a set of samples run in succession than for absolute values, especially for measurements made at different times. Likewise, ratios give greater accuracy since they are less sensitive to any variations in beam current normalization, sample shape effects, etc.

CATALOGUE

Element concentrations from PIXE-PIGME measurements or from other sources can be filed for intercomparison and for reference in artefact studies. Individual measurements are included, as well as averaged or evaluated results, with appropriate file numbers for identification. Information can then be derived on the extent of sample variability for each source, after allowance for the experimental variability applicable to each measurement. Space is also available in the catalogue for element ratios.

The catalogue file is used for the preparation of plots and listings and also as a source for cluster analysis and discriminant function analysis. This permits tests of the significance of differences in multi-element results from sources with different provenance, and the comparison of artefact results with source groups of interest in a particular study. However, such comparisons can be made with greater accuracy if sources and artefacts are included in the one series of runs rather than being measured at different times.

PACIFIC OBSIDIAN

A review of the geographic distribution and characterization studies of volcanic glasses from the Pacific region has been given by Smith, Ward and Ambrose (1977) who summarize reported data for 19 types of obsidian. They include sources from locations ranging from Easter Island through to Indonesia, plus one source from Hawaii. Their results are grouped into Oceanic and circum-Oceanic glasses on the basis of the molecular proportion of soda plus potash relative to alumina. However, in this grouping it is necessary to classify obsidian from Mayor Island (New Zealand) and the eastern end of Fergusson Island (Papua New Guinea) as Oceanic.

In the present study, the same suite of samples has been used plus others from the Solomons, New Zealand, New Guinea and Australia. This set contains all known sources from the South Pacific which have adequate provenance information. Several other types of obsidian composition have been observed during the study of artefacts and these further enlarge the catalogue of types to be considered in future studies.

Consistency

General agreement has been found for the concentrations of a number of elements as determined by the PIXE-PIGME method and as reported by other authors using XRF (Reeves and Ward, 1976) or NAA (Leach and Warren, 1981 and private communication). However, differences of up to 20 to 50% are observed in particular cases, and further work is necessary to improve the overall level of agreement. It is interesting to note that a similar level of agreement is observed with published results of measurements on other volcanic rocks, such as rhyolite (Johnson et al., 1978), when they are from the same region as the obsidian samples (Table 3.1). The PIXE-PIGME method should, therefore, be useful in more general geochemical studies.

Composition Variations

Even when data are available for 20 or more elements, some source material cannot be distinguished separately. This usually occurs for various flows in one volcanic area such as the Willaumez Peninsula in New Britain. On the other hand, obsidian of different ages but from the same area (such as are found at the eastern end of Fergusson Island) may show quite different compositions.

Some of the samples studied show most unusual values for the concentration of some elements. For example, material from Santa Cruz (Solomons) has such high levels of elements such as Ti, Zr and Nb that it is quite unique in composition. Some of the interesting features of the results obtained so far in this study are summarized in Table 3.2

Geochemistry

The ratio of molecular proportions of $Na_2O$ plus $K_2O$ to $Al_2O_3$ is plotted in Figure 3.4 and listed in Table 3.3. The results show a predominance of values near a ratio of 0.9 arising from the number of similar obsidians from New Zealand, New Hebrides, New Britain, etc. Only three sources of Oceanic obsidian are available (Pitcairn and Easter Islands and Puuwaawaa, Hawaii). While these give similar ratios (close to 1.10) they fall within the spread of values from other sources. Higher values are observed for a number of sources from New Zealand and eastern Fergusson Island—as reported by Smith, Ward and Ambrose (1977).

No clear distinction is observed between Oceanic and circum-Oceanic obsidian and, with the measured concentrations varying by up to 30% according to different authors, such a distinction is quite difficult to make. Further work to improve the absolute accuracy may help to provide better information on the geochemistry of obsidian sources.

CONCLUSION

Measurements have been made on all available types of obsidian from the South Pacific region, and the data provide a considerable amount of information which will be made available for use in future studies. Improved accuracy for at least 20 elements is expected when full analysis of the data has been completed.

ACKNOWLEDGEMENTS

We wish to acknowledge the contributions of R. Cawley, M. Scott, D. D. Cohen, L. Russell and E. Clayton to the methods used in this work, and the support of the staff of the Lucas Heights 3 MeV accelerator, as well as of the Australian Institute of Nuclear Science and Engineering.

Table 3.1 Comparison of PIXE-PIGME and geochemical results for Admiralty region obsidian

| | Pam | | Lakou | | Umrei Solang | |
|---|---|---|---|---|---|---|
| Element | JST78** | AAEC | JST78 | AAEC | JST78 | AAEC |
| Na | 3.51% | 3.370% | 3.87% | 3.890% | 3.71%* | 3.640% |
| Al | 7.04% | 6.970% | 7.56% | 6.990% | 7.33% | 7.360% |
| Si | 34.39% | 38.370% | 32.95% | 33.320% | 33.19% | 33.040% |
| K | 3.54% | 4.460% | 3.18% | 3.580% | 3.29% | 3.780% |
| Ca | 0.63% | 0.755% | 1.06% | 1.150% | 0.86% | 1.030% |
| Ti | 0.15% | 0.151% | 0.22% | 0.257% | 0.18% | 0.210% |
| Mn | 0.05% | 0.049% | 0.06% | 0.064% | 0.05% | 0.052% |
| Fe | 1.48% | 1.610% | 2.25% | 2.300% | 1.70%* | 2.020% |
| Zn | 49 | 63 | 54 | 133 | 55 | 71 |
| Rb | 153 | 177 | 134 | 155 | 141 | 172 |
| Sr | 47 | 55 | 93 | 86 | 72 | 83 |
| Y | 38 | 27 | 40 | 22 | 39 | 38 |
| Zr | 302 | 266 | 499 | 459 | 432 | 423 |
| Nb | 43 | 52 | 50 | 49 | 46 | 35 |
| Ba | 813 | 664 | 821 | 588 | 812 | 640 |
| Pb | 8 | 5 | 9 | 2 | 8 | 5 |
| U* | 3.4 | 3.2 | 2.7 | 2.5 | 3 | 3.9 |

Results in $\mu g\ g^{-1}$ if not labelled %

** JST78 - Johnson, Smith and Taylor (1978)

*AAEC results from Wall (1976)

Table 3.2 Some features of Pacific Obsidian composition

| | Origin | Very Low | Low | High | Very High |
|---|---|---|---|---|---|
| Admiralty | Pam | | | Al | |
| | Solang | | Rb | Sr | |
| | Umrei | | | | |
| | Lakou | | | | |
| | Tuluman | | | | |
| Easter Is. | Te Mamavai | | Rb | F, Y, Nb | |
| | Moto Itu | | | F, Y, Nb | |
| | Maunga Orito | | | F, Y, Zr | |
| Australia | Grampians | | Na | K | |
| East Fergusson | Aiasuna (young) | | Ca, Sr | Zr | |
| | Aiasuna (old) | | Al, Ca, Sr | F, Y, Zr | |
| | Numanuma N. | | Ca, Sr | F, Y, Zr | |
| Indonesia | Leles | | Na | Ti | |
| | Gunung Kiamis | | Na | Ti | |
| Kermadec Is. | Kermadec | Rb | K, Zr | Al, Ca | |
| Hawaii | Puuwaawaa | | | Na, Al, Mn, Zr, Nb | |
| Pitcairn Is. | Pitcairn | | | Al, Nb | |
| New Britain | Pilu | | Rb | | |
| | Voganakai | | Na, Rb | | |
| | Talasea | | Rb | Sr | |
| | Garua | | Na, Rb | | |
| | Waisisi | | Rb | | |
| New Zealand | Weta | | Mn, Sr | | F, Rb |
| | Kaeo (Waiare, Pungaere) | | Al, Sr | Cl, Zn, Y, Zr, Nb | F, Rb |
| | Huruiki | | Ti | | |
| | Fanal Is. | | Mn | | |
| | Awana | | Ti | Al | |
| | Te Ahumata | | Ti, Sr | | |
| | Cook's Bay | | | | |
| | Purangi | | | | |
| | Hahei | | Ti | | |
| | Tairua | | | | |
| | Waihi (red and black) | | F, Y | | |
| | Mayor Is. (Honey, green and black) | | Al, Ca, Sr | S, Y, Zr | |
| | Rotorua (red) | | Ti | | |
| | Rotorua (black) | | Ti | | |
| | Maraetai (red) | | Y | | |
| | Maraetai (black) | | | | |
| New Zealand (cont.) | Ongaroto | | Y | | |
| | Taupo | | Y | | |
| | Banks | | | Ti, Mn, Fe, Sr, Nb | Al |
| | Otago | Na | | Fe, Sr | Ca, Mn |
| | Tamaka | | Y | Ca, S, Mn, Fe, Nb | F, Ti, Sr |

Table 3.3 Molecular ratio $(Na_2O + K_2O)/Al_2O_3$

| Source Region | No. of Sources Within the Region | Ratio |
|---|---|---|
| Australia | 1 | 0.83 |
| Admiralty Is. | 6 | 0.91 |
| Indonesia | 2 | 0.90 |
| New Britain | 10 | 0.83 |
| New Hebrides | 4 | 0.93 |
| New Zealand | 17 | 0.89 |
| East Fergusson | 3 | 1.32 |
| West Fergusson | 2 | 0.98 |
| N.Z. Kaeo | 2 | 1.57 |
| N.Z.Mayor Is. | 3 | 1.54 |
| Marquesas Tamaka | 1 | 1.40 |
| N.Z.Otago | 1 | 0.41 |
| Kermadec Is. | 1 | 0.48 |
| Solomons, Santa Cruz | 1 | 0.23 |
| Solomons, Tikopia | 1 | 0.64 |
| Tonga, Tafahi | 1 | 0.66 |
| Pitcairn | 1 | 1.08 |
| Hawaii, Puuwaawaa | 1 | 1.20 |
| Easter Is., Te Mamavai | 1 | 1.08 |
| Easter Is., Maunga Orito | 1 | 1.26 |
| Easter Is., Motu Iti | 1 | 1.06 |

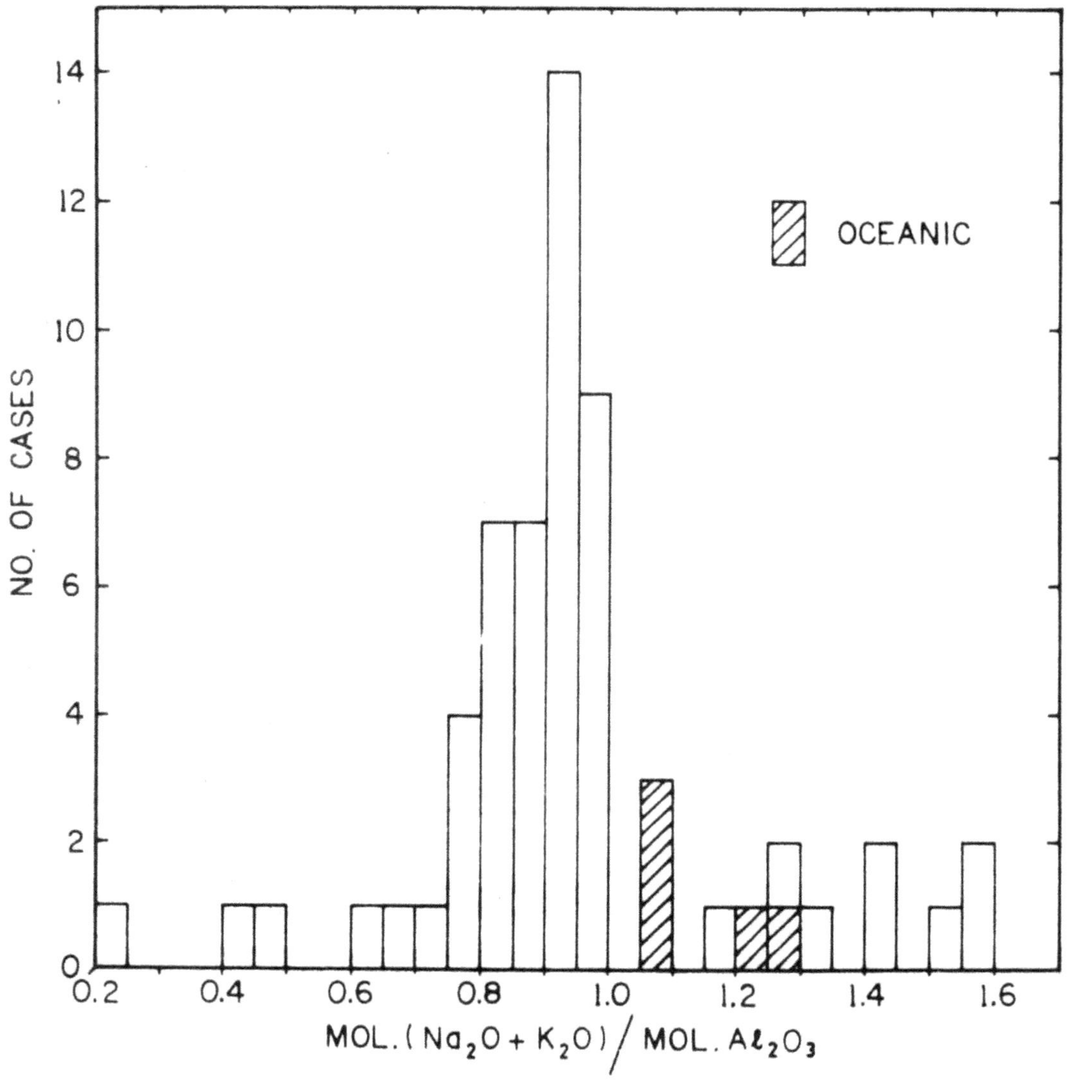


Figure 3.4 Molecular ratio [$(Na_2O + K_2O)/Al_2O_3$]

REFERENCES


Barnard, N. (Ed.), 1976. The proceedings of a symposium on scientific methods of research in the study of ancient Chinese bronzes and South-east Asian metal and other artefacts. National Gallery of Victoria, Australia.

Bird, J. R. and Russell, L. H., 1976. Applications of prompt nuclear analysis techniques to the study of artifacts including southwest Pacific obsidian. Pp. 317-336 In: Barnard, 1976 q.v.

Bird, J. R., Russell, L. H., Scott, M. D. and Ambrose, W. R., 1978. Obsidian characterization with elemental analysis by proton induced gamma-ray emission. Analytical Chemistry 50(14): 2082-2084.

Duerden, P., Cohen, D. D., Clayton, E., Bird, J. R., Ambrose, W. R. and Leach. B. F., 1979. Elemental analysis of thick obsidian samples by proton induced X-ray emission spectrometry. Analytical Chemistry 51(14): 2350-2354.

Duerden, P., Bird, J. R., Scott, M. D., Clayton, E., Russell, L. H. and Cohen, D. D., 1980. PIXE-PIGME studies of artefacts. Nuclear Instruments and Methods 168: 447-452.

Johnson, R. W., Smith, I. E. M. and Taylor, S. R., 1978. Hot-spot volcanism in St Andrew Strait, Papua New Guinea: geochemistry of a Quaternary bimodal rock suite. BMR Journal of Australian Geology and Geophysics 3(1): 55-69.

Leach, B. F. and Warren, S. E., 1981. Neutron activation analysis of New Zealand and Oceanic obsidians: towards a simple screening technique. Pp 151-166 In: Leach, B. F. and Davidson J. M. (Eds.), Archaeological Studies of Pacific Stone Resources. British Archaeological Reports, International Series 62.

McCallum, G. J., McFadgen, B. G., Moore, P. R., Richardson, L. M. and Sheppard, R. A., 1979. Characterization of New Zealand obsidian by energy-dispersive X-ray fluorescence. Manuscript 1001, Institute of Nuclear Sciences, D.S.I.R., Wellington.

Nielson, K. K., Hill, M. W., Mangelson, N. F. and Nelson, F. W., 1976. Elemental analysis of obsidian artifacts by proton particle-induced X-ray emission. Analytical Chemistry 48(13): 1947-1950.

Reeves, R. D. and Ward, G. K., 1976. Characterization studies of New Zealand obsidians: towards a regional prehistory. Pp. 259-287 In: Taylor, 1976 q.v.

Smith, I. E. M., Ward, G. K. and Ambrose, W. R., 1977. Geographic distribution and the characterization of volcanic glasses in Oceania. Archaeology and Physical Anthropology in Oceania 12(3): 173-201.

Taylor, R. E. (Ed.), 1976. Advances in obsidian glass studies. Noyes Press.

Wall, T., 1976. Use of the research reactor MOATA and associated facilities for the source identification of obsidian artifacts. Pp. 337-350 In: Barnard, 1976 q.v.

# 4. SOME ASPECTS OF THE STONE INDUSTRY OF THE ANCIENT INHABITANTS OF SOUTH PRIMORYE, USSR

T. P. Butylina

## INTRODUCTION

For primitive man, stone served as the main material for making tools, ornaments, etc. Only stones that met certain technological requirements were used, and they were deliberately sought out during the Palaeolithic, and became the object of mining during the Neolithic.

According to Petrun, the primitive man used 100 kinds of rocks and minerals of various properties. In one case—nephrite—he took advantage of the hardness and toughness of the stone, in another—flint and shale—he used its ability to flake conchoidally, in yet a third case—jasper—he utilized its homogeneity and strength. Flint was widely used by primitive man in making tools, whereas limestone and coaly shale, although similar to flint in colour and structure, were of no use for this purpose. Alabaster, calcite and quartz, all white minerals of similar colour, were used differently: the soft alabaster was used for small artefacts (ornaments and utensils), while calcite which has high cleavage, and quartz which is hard, were used for making cutting or chopping tools. Semenov (1968) states that the choice of a certain rock corresponded to a definite stage in the development of stone processing techniques. As technology developed, the choice of rocks widened. In the Palaeolithic, for example, stone processing techniques consistedof breaking off, splitting, and therefore flint, quartz and obsidian were the main rocks used. When these materials were not available, man used quartzite, siliceous tuff, felsite, siltstone, horny amphibolite, andesite, and other rocks. Techniques such as plishing and pecking, which were widely used in the Neolithic, significantly enlarged the range of rocks chosen for tool making. Such rocks as basalt, diorite, diabase, nephrite etc. began to be worked too. These techniques of stone working favoured the introduction of relatively soft rocks (such as slates) into use. Polishing transformed then into axes, knives and adzes which suited the manufacture of boats by hollowing cut tree trunks, and also the building of houses. It has been shown experimentally by Semenov (1968) that different rocks need different techniques of working. For example, the flaking methods appropriate to soft rocks (such as shales, limestones etc.) differ significantly from those most suited to working hard rocks (such as flint, quartz etc.). An archaeologist may obtain interesting data if he considers stone artefacts as mineral formations, and determines their physical and mechanical properties.

## STONE ARTEFACTS OF SOUTH PRIMORYE

Petrographic studies of stone artefacts from archaeological sites of the Neolithic and early Iron ages in different areas of South Primorye have enabled us to judge what properties served as the basis for choosing raw materials for making different tools, as well as determining the sources of the materials. Functionally, the tools considered can be divided into four groups as follows.

### Tools for Making Tools

The rocks used for this purpose are characterized by extreme homogeneity, hardness and strength. Such rocks as jasper and others with similar properties were used for making tools of many different functions. Semenov found that when a hammer stone 'pick' made of flinty rock was used, the blow was sharper, and should have facilitated better chipping; however, the quality of flakes actually deteriorated. Flakes were shorter and more curved. On the other hand, a 'pick' made of softer rock with the surface crushed and pitted by striking (for example andesite or porphyrite) produced larger a and more rectilinear flakes.

### Chopping Tools

The rocks for these tools are characterized by relative hardness, well-defined lamination, and some toughness. They are thus suitable for the sawing, polishing, and sharpening techniques employed in making these tools. For chopping tools, the choice of rocks is fairly wide: horny aleurilites, slates, hornfels, amphibolites. Along with hard materials, some softer rocks were also used (for example sericitic quartz schist). Their ability to take a polish made them well suited to the manufacture of axes and adzes.

The choice of rocks for axes and adzes was similar. Most polished tools were made from amphibolite, which has the highest toughness. The laminations of aleurolites meant that they could be well sharpened. This characteristic is especially valuable for making knives.

### Agricultural tools (pestles and hoes)

Petrographic studies of thin sections showed that agricultural tools in South Primorye were made from sedimentary and sedimentary-metamorphic rocks. Such tools as pestles, requiring greater hardness and strength, were made of fine-grained feldspathic quartz sandstone. In their hardness, the rocks used for making pestles are close to the 1-group rocks (picks). To manufacture hoes, the prehistoric inhabitants used rocks characterized by laminations, as well as strength and hardness. Aleurolites of varying degrees of horniness met these demands; their density approached that of rocks used for chopping tools.

### Small tools such as arrow heads

Petrographic studies have revealed that arrow heads were made of fresh metamorphic rocks, such as chlorite-actinolite and sericite schists. These rocks are softer, and have perfect cleavage. These properties allowed thin plates to be split off easily and quickly. From these, arrows of definite shape, with clear-cut faces could be manufactured, and polished easily and quickly. The petrographic data on different tools has indicated that during manufacture the choice of rocks was determined by the ultimate purpose of the tools. In addition, the manufacturing technique employed (sawing, polishing, drilling, striking) depends on such properties as hardness, viscosity, strength and lamination.

## CONCLUSION

Thus, the correctly chosen stone material, meeting the article's purpose, and its method of treatment, improved the labour productivity of ancient man. Semenov has established experimentally that 2.5-3 hours was required to make polished axes of relatively soft and viscous rocks (such as amphibolite); while in the case of rocks similar to granites and diorites in hardness, an axe may have required polishing for 12-15 hours.

The main features of the stone industry of ancient man are closely dependent on the raw material available; that is, they were conditioned by the geological structure of man's habitat. After comparing the petrographical character of stone implements and the distribution of the rocks in a given area, one may conclude that the ancient inhabitants of South Primorye made tools suiting different functions by taking advantage of the particular characteristics of the raw materials available.

## REFERENCES

Semenov, S. A., 1968. The development of technology in the Stone Age. Nauka, Leningrad.

## 5. НЕКОТОРЫЕ АСПЕКТЫ ИЗУЧЕНИЯ КАМЕННОЙ ИНДУСТРИИ ДРЕВНИХ ОБИТАТЕЛЕЙ ЮЖНОГО ПРИМОРЬЯ

Т.П. Бутылина

Тихоокеанский институт географии ДВНЦ АН СССР

Владивосток, СССР

Камень служил первобытному человеку основным материалом для изготовления орудий труда, украшений и других целей. Первобытный человек обрабатывал не первые попавшиеся под руку камни, а только такие, которые отвечали определенным технологическим требованиям и являлись (палеолит) объектами целеустремленных поисков, а в дальнейшем и добычей этих первых полезных ископаемых (неолит).

По данным В.Ф. Петруня (1959) первобытный человек использовал для изготовления различных предметов около ста горных пород и минералов, обладающих определенными свойствами. В одних случаях использовались твердость и вязкость (нефрит); в других - способность принимать от удара форму (кремень, сланец); в третьих - однородность и прочность (яшма).

Так например, кремень широко применяется для изготовления орудий первобытного человека, а сходные с ним породы по цвету и структуре (чистый известняк, чистый углистый сланец) для этой цели непригодны. Сходные по цвету (белые) минералы - алебастр, кальцит, кварц применялись также по-разному: мягкий алебастр - в качестве материала для мелких поделок (украшения, бытовые предметы), а кальцит, обладающий высокой спайностью, и твердый кварц применялись для режущих или ударных орудий.

Работами С.А. Семенова (1968) было установлено, что определенному этапу развития техники обработки камня соответствовал определенный выбор пород.

В процессе эволюции техники выбор пород расширялся. Так, в палеолите основными способами обработки камня были оббивка, скалывание, расщепление и ретуширование материалов. Для этого необходимы были породы, обладающие высокой твердостью и четко выраженной отдельностью (следовательно чувствительные и на удар). Основными породами, из которых изготовлялись палеолитические орудия были кремень, кварц, обсидиан. В местах, где его не было, человек применял кварцит, кремнистый туф, фельзит, алевролит, ороговикованный амфиболит, андезит и др.породы.

Техника пиления, шлифования, точечно-ударная, получившие наиболее широкое распространение в неолите, существенно расширили выбор пород камня. Были введены в эксплуатацию такие породы, как базальт, диорит, диабаа, нефрит и др.

Эта техника обработки камня позволила также звести в употребление довольно мягкие породы (сланцы).

Шлифование превращало их в топоры, ножи, тесла, пригодные для долбления лодок, строительства жилищ.

С.А. Семенов (1968) опытным путем установил, что различные породы камня при обработке их одним и тем же способом требовали различных технических приемов. Например, ударный способ обработки мягких пород (сланцев, известняков) существенно отличался от ударного способа обработки твердых (кремня, кварца и др.).

Рассматривая каменные излелия, как минеральные образования и определяя физические и механические свойства, археолог может получить ряд интересных данных.

На основании петрографических определений минерала или горной породы в каменных изделиях из археологических памятников эпох неолита и раннего железа в различных районах Южного Приморья можно судить по каким свойствам древний человек выбирал сырье для изготовления различных орудий, выяснить некоторые технологические приемы его обработки, а также установить происхождение этого сырья.

Все рассмотренные орудия по их функциональному назначению можно разделить на четыре группы.

Для орудий первой группы - орудия для производства орудий - подбирались породы, отличающиеся исключительной однородностью, твердостью, прочностью. Такие породы, как яшма и другие с такими же свойствами, человек использовал для изготовления орудий различного назначения. С.А. Семеновым было установлено, что при употреблении каменного отбойника из кремнистой породы удар отличался большей резкостью, облегчающей скалывание, но качество отщепов ухудшалось. Отщепы получались укороченных пропорций и с большей кривизной профиля.

Отбойники из более мягких пород, поверхностный слой которых дробился и сминался от удара (например, андезитовый порфирит), давали отщепы более длинные и более прямые в профиле.

В группе второй рубящих орудий - представлены породы, обладающие относительно высокой твердостью, отчетливой слоистостью и некоторой вязкостью, т.к. изготовление данных орудий производилось путем пиления, шлифования, заточки. Выбор пород для рубящих орудий довольно многообразен: алевролиты ороговикованные, сланцы, роговик, амфиболиты. Наряду с твердыми породами использовались и такие мягкие породы как сланец кварц-серицитовый. Шлифование делало их пригодными для изготовления топоров, тесал. Выбор породы для топоров и тесал был один и тот же, причем наиболее отшлифованные орудия изготовлены из амфиболита, обладающего наибольшей вязкостью. Слоистость алевролитов способствовала хорошей заточке орудий, что особенно ценно при изготовлении ножей.

III группа. Земледельческие орудия (куранты, мотыги).

Петрографическое изучение шлифов показало, что земледельческие орудия представлены осадочными и осадочно-метаморфическими свежими породами. Такие орудия, как куранты, требующие большей прочности и твердости, изготовлены из мелкозернистого полевошпатового кварцевого песчаника. По твердости породы, служившие для изготовления курантов, близки породам I группы орудий (отбойники). Для изготовления мотыг употреблялись породы, обладающие, кроме прочности, и твердости, слоистостью - этим требованиям удовлетворяли алевролиты в различной степени ороговикованные, приближающиеся по твердости к породам, из которых изготовлялись рубящие орудия.

IV группа. Мелкие орудия (наконечники стрел).

Петрографическое изучение показало, что наконечники стрел изготовлены из свежих метаморфических пород: хлорит актинолитовых и серицитовых сланцев. Эти породы обладают повышенной мягкостью, совершенной спайностью. Эти свойства материала давали возможность легко и быстро откалывать тонкие пластины, изготавливать из них стрелки определенных форм с четкими гранями, а затем также легко и быстро шлифовать изделие.

Приведенные материалы по петрографическому изучению различных орудий показывают, что при их изготовлении выбор пород определялся назначением орудий и техникой их изготовления (пилением, шлифованием, сверлением, ударной), что, в свою очередь, обусловливалось такими свойствами как твердость, вязкость, прочность, слоистость.

Таким образом, правильно выбранный каменный материал, отвечающий назначению предмета, приемам его обработки, повышал производительность труда древнего человека. Опытами С.А. Семенова было установлено, что для изготовления шлифованных топоров из относительно мягких, вязких пород (амфиболит) требовалось 2,5-3 часа работы, а из пород, приближающихся по твердости к гранитам или диоритам, топор может быть отшлифован за 12-15 часов.

Основные особенности каменной индустрии древнего человека тесно связаны с сырьем, которым человек располагал для изготовления своих орудий, т.е. обусловливались геологическим строением района обитания древнего человека. Сравнивая петрографические определения каменного инвентаря и распространение этих пород в данном районе, можно сделать вывод, что древние жители Южного Приморья изготавливали орудия различного назначения из местного сырья.

ЛИТЕРАТУРА

СЕМЕНОВА,С.А.,1968. РАЗВИТИЕ ТЕХНИКИ В КАМЕННОМ ВЕКЕ. НАУКА, ЛЕНИНГРАДЕ.

# 6. MODELS OF PHILIPPINE PREHISTORY: A REVIEW OF THE FLAKED STONE INDUSTRIES

P. J. F. Coutts and J. P. Wesson

## INTRODUCTION

During the past twenty years a number of important archaeological discoveries have been made in the Philippines, principally by Robert Fox and his co-workers. The foci of much of this research have been caves on the island of Palawan, surface sites in the Cagayan Valley in northern Luzon, and caves in Albay and Sorsogon Provinces in southern Luzon. The stone assemblages from these sites span a large time range, commencing about 30,000 years B.P., and following each new discovery Fox and his colleagues have endeavoured to order their archaeological data within fairly flexible interpretative models.

In this paper the utility of these models is examined, and special prominence is given to the role of stone tool technology in them. Stone artefacts have been singled out because they are found in archaeological sites dating from the earliest periods of Philippine prehistory and can therefore provide a monitor of change for one aspect of material culture. Having assessed current models, an alternative approach is suggested.

## MODELS IN PHILPPINE PREHISTORY

Up until the Second World War Philippine archaeology was dominated by foreign field workers, particularly H. Otley Beyer. Beyer was a tireless field worker but over a number of decades, failed to publish a single site report or methodology paper. Consequently many of his basic data remain unpublished today.

Beyer was influenced by models that were fashionable in his day and used them to order his archaeological materials. Under the influence of Heine-Geldern he proceeded to characterize Philippine prehistory within the framework of chrono-technological and diffusionist models. These models were built on traditional beliefs such as those relating to the settlement of the Philippines by peoples from nearby South East Asia, the presence and/or absence of type fossils in the archaeological record, and cross-cultural comparisons. Evangelista (1967) has discussed the merits of Beyer's theses in some detail and convincingly demonstrates that they are unsound and inappropriate for Philippine prehistory.

Robert Fox, an American who was head of the Anthropology Division of the National Museum of the Philippines until five years ago, has been the most influential person in Philippine prehistory in the post-war years. He has been associated with a number of important archaeological projects and has often been the senior researcher. His projects include studies of the Tasaday

(Fox, 1976), archaeological investigations in the Cagayan Valley (Fox et al., 1972; Fox, 1973; Fox and Peralta, 1974) in Sorsogon and Albay Provinces (Fox and Evangelista, 1957a, 1957b), Batangas (Fox, 1959) and Palawan (Fox, 1962; 1967a; 1968; 1970; 1973). Fox has been seriously ill in recent years and his oft stated intention to publish full accounts of some of these important excavations has been thwarted.

Fox has stated that he is acutely aware of the limitations of the archaeological evidence in the Philippines (Fox, 1970: 124) and of the shortcomings of extant interpretative models, but despite the fact that he has strongly advocated data collection to build regional sequences (see Fox, 1970: 1, 4, 19; 1967a: 93, 113) he has been a major patron of the "national sequences" or all embracing "explanatory models" syndrome. He has not been beyond reasoned speculation either (e.g. Fox, 1970: 94, 98, 134, 156, 160-162. 164, 166) and in this respect is in good company with Beyer.

What has tended to happen in the Philippines (as has happened in Australia and elsewhere) is that when an important site is discovered it suddenly becomes the key site (because important sites are rare) and the data from the site are used to define the beginning and end of sequences and the arrivals and departures of important fossil types for the whole country. Thus, according to Fox, the Iron Age commences synchronously with the earliest radiocarbon dates obtained for the Iron Age on Palawan (approximately 2,200 B.P.). In fact, the Palawan sites have become the focus of discussion for all major events in Philippine prehistory.

The Palawan research programme has been conducted methodically and has revealed a number of rich sites with long histories. For these reasons alone the research is important. However, rather than treating the sites as sources of information to be used to elucidate specific and local aspects of prehistoric settlement patterns (used here in the ecological sense after Hole and Heizer, 1973: 355), Fox has chosen to argue that they have a special significance which in turn allows him to generalize from his sequences over the whole of the Philippines. He has generally glossed over the question of regional diversity.

Fox uses an essentially techno-developmental model to order his archaeological data and relies strongly on the presence or absence of fossil markers to define developmental stages. He orders Philippine prehistory from an Upper Palaeolithic (ending about 7,000 B.P.) followed by an Early Neolithic (7,000 - 3500 B.P.), Late Neolithic (3,500 - 2,500 B.P.),Early Metal Age (2,500 - 2,200 B.P.), Developed Metal Age (2,200 - 1,100 B.P.), Age of Contact (1,100 - 400 B.P.) to a Colonial Phase (post-400 B.P.). Each phase is characterized by the eclipse or arrival of cultural traits, technologies or economic modes.

Essentially this model is a more elaborate version of the one used by Beyer, workable on a regional basis but unsuitable for ordering data over wider areas. Inherent in Fox's model are a number of assumptions: that all or most new items of material culture that are of social and economic importance are imported (e.g. jar burials), that settlement of the Philippines took place at different times by small groups of people moving through the archipelago in boats, and that changes in stone tool technology will reliably reflect or herald other cultural changes.

This model suffers from the same weakness as Beyer's; it ignores regional diversity (in technology, geography and geology etc.). However, what is worse, it does not consider alternative explanations and plays down the fact that the archaeological information gives only select information about prehistoric settlement patterns. Thus, technologies may vary through association with different ranges of activities, which in turn may reflect such factors as seasonality and specialization. They may only be aspects of a settlement pattern shared by a culturally homogeneous group of people. Nor was it necessary for new cultural traits to be brought to an area by immigrants; they could arrive through trade either via visiting merchants or by way of locals going abroad. In the Palawan region, one major stone industry persists over a long period of time and, if anything, at the techno-complex level this suggests continuity of a local culture.

Wilhelm Solheim, another American, has greatly influenced current perspectives on Philippine pottery traditions. His studies have concentrated on the Philippine Iron Age, and in particular, upon the type, variety and distribution of earthenware pottery (Solheim, 1955; 1958; 1959b; 1964a; 1964b). The focus of his early work was the Guthe collection of pottery, mostly surface finds, gathered during the 1920s. He supplemented that study by the excavation and analysis of largely disturbed, burial complexes on the island of Masbate, and with observations of extant pottery making traditions in several areas of the Philippines, the Pacific Islands and South East Asia (Solheim, 1952a; 1952b; 1952c; 1954; Solheim and Shuler, 1959). These studies led Solheim to divide Philippine earthenware into three major complexes—the Kalanay, Bau and Novaliches—on the basis of vessel form and types of decoration. His studies have been extended to include pottery traditions on the Asian mainland and he has postulated a large family of related pottery (his Sa-Huynh-Kalanay Complex) occurring throughout South East Asia and the Pacific (e.g. the works cited above and Solheim 1959a; 1967).

At the time he was working, Solheim's arguments were innovative, and today his ideas still form a useful framework for establishing very general relationships between pottery complexes in the Philippines and elsewhere. However, the Philippine archaeological evidence has become increasingly complex, especially for the "Neolithic", and it is clear that there is now an urgent need to refine Solheim's model and use well documented archaeological evidence from a much greater range of sites (since most of Solheim's data derive from burial/funerary contexts). There is a need to test his cross-dating techniques by using only well-provenanced data, and to consider and compare total assemblages of artefacts rather than specific items such as pottery.

Alfredo Evangelista, the Assistant Director of the National Museum of the Philippines, is the most senior Filipino archaeologist and has worked with Beyer, Fox and Solheim. His review papers (1962; 1967) are penetrating documents which clearly express disenchantment with the directions of Philippine archaeology in recent times. He has resisted the temptation to synthesize Philippine prehistory at this stage in its development.

F. Lando Jocano (1975a) has made the most recent attempt to synthesize Philippine prehistory. He is concerned with the public image of the discipline

(1975a: 1), and in a book designed for a general readership has produced a document with nationalistic if not political overtones. Unlike Fox, he attempts to identify and highlight indigenous elements of Philippine culture which have their roots in prehistory and which are the result of Filipino innovation, and to play down traits that may have come from South East Asia.

Jocano also recognizes the dangers of synthesizing Philippine prehistory at this time (1975a: 3), but justifies his own efforts by noting that "few studies in prehistory are ever written on the basis of satisfactory traces of empirical evidence" (ibid.). The book is a useful document because it provides a summary of relevant archaeological evidence in a single volume; but his main contribution is his attempt to present the data in an innovative interpretative framework. He used (1967: 139; 1975a) a multi-linear developmental model with five stages (after the scheme of Willey and Phillips, 1958) which works on the premise that "any culture can emerge, develop and predominate in any phase, without necessarily passing through the preceding one" (Jocano, 1967: 139-140; 1975a).

In theory, as Evangelista (1967: 67-68) has conceded, this model is potentially far more useful than any of those which have been used in the past, but unfortunately the data necessary to describe his model, in detail, are not yet available. Indeed, Jocano does not attempt to evaluate critically the data which he uses. Thus, both popular and hackneyed interpretations emerge and are not tested. This is to some extent forgivable, as any attempt to treat the data critically and on a regional basis would have resulted in an incoherent and unreadable account of Philippine prehistory. However, the basic fault of the book remains: excessive generalization from the data.

SOURCES OF DATA

In order to utilize viable models for ordering and explaining prehistory, whether they be local, regional or universal models, it is necessary to have well provenanced archaeological data. Ideally these should derive from "type" sites in different environmental locations, and reflect the different settlement patterns and seasonal occupations of the prehistoric inhabitants through time. It is also important to have data which describe both present and past environments.

Although much environmental information is available in published works, detailed studies which are relevant to areas of archaeological interest are generally rare. In fact at present there is not a single archaeological site or regional report that sets a site or sites in proper environmental perspective.

There is no shortage of archaeological sites in the Philippines; thousands have been found. However many have not been reported in detail and many have been looted (see Beyer, 1947). In general, systematic site survey has been confined to cave sites, since sites situated in other settings (called here "open sites") are difficult to find. Open sites tend to be screened by vegetation, they have been destroyed or damaged by development, and their locations are kept secret by profit motivated treasure hunters. Further, as Spoehr (1973: 30) points out, the tendency of prehistoric slash and burn agriculturalists to shift regularly meant that their rubbish deposits were often not large and may not have marked the landscape sufficiently for archaeologists to recognize them easily.

In practice, open sites are generally found by accident; for example when farmers turn up porcelains during ploughing. Few have in situ deposits that can be systematically investigated (e.g. the Cagayan Valley sites) and many are burial sites dating to the historic period (from about 1,000 B.P.) and are of little interest to the prehistorian. Indeed, the majority of recorded open sites in the Philippines are historical sites.

Since the chances of finding in situ open sites are poor, Philippine archaeologists have tended to concentrate on the exploration of caves. However not all caves were used by prehistoric man, and of those which contain cultural debris most have been used as burial places (cf. Fox, 1970: 161). Thus the dozens of cave sites discovered on Masbate (Solheim, 1955: 1959b), in Sorsogon and Albay Provinces of Luzon (Fox and Evangelista, 1957a; 1957b) and on Palawan (Fox, 1970) are nearly all burial sites. These sites yield only certain ranges of funerary artefacts and thus thrown light on only very select aspects of the prehistoric settlement patterns.

It is never made clear in any of the Philippine reports what the authors regard as "habitation sites". For example, Fox (1970: 22) regards the flaking floors in the Tabon Cave as "habitation levels", but in the same publication (Fox, 1970: 161) he ponders the fact that no "habitation sites" of the jar burial people had been excavated at that time. In neither case does he define what he means by the term "habitation". Cherry (1978: 9) excavated flake tools on Buad Island and although he concluded that it was not a habitation site he does not give his reasons or an alternative functional classification.

The result of twenty years of excavation in the Philippines is a bewildering array of data from a very select range of sites. Moreover, although many of the sites are concentrated together in particular areas, a necessary prerequisite for constructing sound regional sequences (e.g. for Masbate, Batangas etc.), the data are not from a representative sample of sites, and we therefore have only limited knowledge of the prehistoric settlement patterns. One of the most frustrating aspects of Philippine archaeology is the non-publication of site reports. Before the Second World War there were only a few publication outlets, the most important being the Philippine Journal of Science and the newspapers and magazines. During the post-war period the Journal of East Asiatic Studies (now defunct) and, more recently, the Philippine Quaterly of Culture and Society and a number of foreign journals (e.g. Asian Perspectives) have been the main publication forums. Much information is contained in unpublished manuscripts and it is often difficult to obtain copies of relevant papers. Recent publications dealing with Philippine archaeology commonly contain references to a significant number of unpublished manuscripts. For example, in Fox's Tabon Cave report (Fox, 1970) approximately 10% of the references are cited "typescript", and another 25% are difficult to obtain. In this paper also, a number of unpublished reports have been referred to (see Table 6.1). There are other unsatisfactory features of reports published within the Philippines: they tend to be "preliminary" and contain little data and there are many errors and omissions (e.g. Fox, 1970; Jocano, 1975a). Finally, there are difficulties in using the collections of artefacts from surface sites and excavations housed in the National Museum of the Philippines. A number of collections of stone tools were examined for this study, including assemblages from Tabon Cave,

Palawan, Laurente Cave (Penablanca area of northern Luzon), and four surface assemblages from the vicinity of Cabatuan in Iloilo Province, Panay Island. The major problem with using the surface collections is that they are rarely accompanied by field notes so the collection method is unknown. This reduces the research potential of the collections.

After reviewing the Philippine literature, it was decided that there are only about a dozen publications that can be used in model building exercises. There are a number of relevant unpublished sites which could have been used in this exercise. The important Dimolit site on Palawan Bay, for example, has been excavated by Peterson (1974b) but the authors have not had access to his data. The range of information available from the principal published papers is summarized in Table 6.1. The locations of sites mentioned in the text are shown in Figure 6.1.

## SOME PREREQUISITES FOR BUILDING MODELS OF PHILIPPINE PREHISTORY

There are a number of factors which need to be taken into consideration when designing models for Philippine prehistory. Some of these are discussed below.

### The Geographical Position of the Philippines

The Philippine Islands are situated close to the Asian mainland and some of the islands were joined to the mainland during the last glaciation. Thus one might anticipate that cultural and biological influences from South East Asia would be reflected in the archaeological record in the Philippines and, conversely, that elements of Philippine culture may be found in the archaeological records of South East Asia.

### Ecological Diversity

Today the Philippines are made up of 7,083 islands which cover an area of 296,300 square kilometres. However, more than half this area is covered by the two largest islands, Luzon and Mindanao. In spite of the relatively small land area, the archipelago is ecologically diverse; its biological and mineral resources are not evenly distributed (see Dickerson, 1928; Wernstedt and Spencer, 1967). This was likely to have been the case in prehistoric times too, when man's impact on the environment was still minimal. Consequently, we might anticipate a diverse range of aboriginal cultures at every stage in prehistory.

### Cultural Diversity

Today the Philippine Islands are culturally diverse in spite of 400 years of colonial rule. There are still many different ethnic groups and about 70 languages and dialects are spoken (Wernstedt and Spencer, 1967:153).

These differences were more pronounced 400 years ago at the beginning of Spanish rule (cf. Quirino and Garcia, 1958) and we might infer increasing cultural diversity as we go back in time. However many non-material aspects of this cultural diversity will not be reflected in the archaeological record (e.g. language), whilst the material culture aspects may seem to become more uniform as we step back in time. This is an archaeological dilemma.

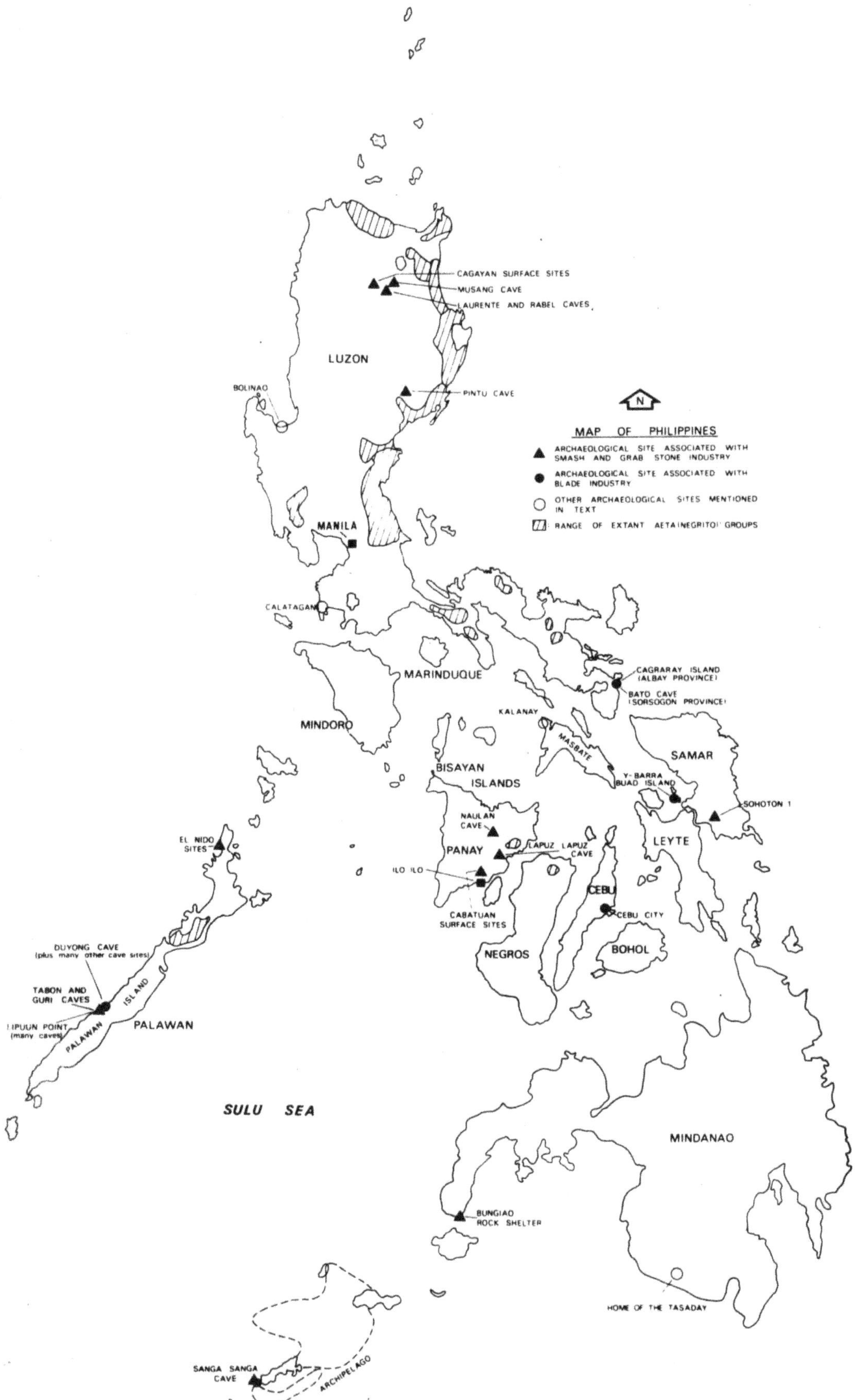


Figure 6.1 Map of the Philippine Islands showing the principal archaeological sites.

TABLE 6.1 Gazetteer of select Philippine archaeological sites associated with stone tool assemblages: Data Classification

| | | | CULTURAL SYSTEM | | | | | | | | | | | | | ENVIRONMENTAL SYSTEM | | | | | | | | |
|---|---|---|---|---|---|---|---|---|---|---|---|---|---|---|---|---|---|---|---|---|---|---|---|---|
| | | | MATERIAL CULTURE SUBSYSTEM | | | | | | | | ECONOMIC SUBSYSTEM | | | RELIGIOUS SUBSYSTEM | | | | | | | | | |
| | | | | Pottery | | | | | | | | | | | | | | | | | | | | |
| | | | | Earthenware | | | Tradewares | | | | | | | Burial Practices | Physiographic Subsystem | | Climatic Subsystem | | Floral Subsystem | | Faunal Subsystem | | Data Classi- fication |
| Site Name | Type of Site | Date Range (BP) | Stone Tools | Utilit. | Grave Goods | Burial Jars | Utilit. | Grave Goods | Burial Jars | Ivory/ Bone Antler Tools | Other | Hunting | Gath- ering | Trade | Rituals etc. | Extant | Preh. | Extant | Preh. | Extant | Preh. | Extant | Preh. | |
| Tabon | Cave Habitation | 8,500-30,000 | E & F | X(P&A) | X | X | | | | | | X | | | | X | X | X | H | X | H | X | X | C |
| Buri Chamber A | " | 4,070 | E ? | ? | X but from Chamber B | | | | | | | | | | None from Chamber A | X | H | X | H | H | H | H | H | C |
| .L/1 | " | 880-1,010 | E & F | X | | | | | | X | Bone & antler | X | X | | | H | H | H | H | H | H | X | H | C |
| A/1 | Rock-shelter Habitation | Unknown | E | X | | | | | | X | Bone tools | X | X | | | H | H | H | H | H | H | H | H | C |
| abatuan* (# sites) | Surface site | Unknown | E | | | | | | | | | | | | | H | H | H | H | H | H | H | H | A |
| ungiao | Rock-shelter Habitation? Burial? | 1,100-1,500 & Earlier | E | X | | | X | X | | | Many found incl. iron & glass | X | X | X | X | H | H | H | H | H | H | H | H | C |
| anga anga | Cave Habitation | 6,700-7,700 | E & F | X | | | | | | X | X incl. shell adzes | | | | | H | H | H | H | H | H | H | H | C |
| ohoton I | Cave Habitation | 4,000-11,000 | E & F | X | | | | | | | | | | | | X | H | X | H | H | H | H | H | C |
| agayan alley* | Surface Sites | Unknown | E & F | | | | | | | ? | | | | | | X | H | X | H | X | H | H | H | C |
| aurente | Cave Habitation | Unknown | E | X | | | | | | | | X | X | | | X | H | X | H | H | H | H | H | A |
| abel | " | Unknown | E | X | | | | | | | | X | ? | | | X | H | X | H | H | H | H | H | A |
| intu | " | 2,300-3,700 | E & F | X | | | | | | X | | X | X | | | H | H | H | H | H | H | H | H | C |
| lusang | " | 4,000-2,000 | E | X | | | | | | ? | X Several | X | X | Pos-sibly | | ? | ? | ? | ? | ? | ? | ? | ? | A |
| orsogon lbay rovince* | Cave Burials | 300-600- | G ? | X | X | X | X | X | X | | X many | | | | X | X | H | X | H | H | H | H | H | C |
| uyong | Cave Habitation | 4,600-? 7,000 | G | ? | X many | X | | | | | shell adze 4,630 BP & other goods | | | | X | X | H | X | H | X | H | H | H | C |
| arrio -Barra | Surface Habitation? | Unknown | G | | | | | | | | | | | | | X | H | X | H | H | H | H | H | B |

KEY:

A - Unpublished data.

B - Fully published.

C - Preliminary report only.

H - No data available at present.

E - Smash and Grab Industry.

F - "Cobble" tools present.

G - Blade Industry.

X - Present/some data available.

* - Several sites in these areas.

## Implications from Ethnographic and Historical Data

It is often possible to describe aspects of regional cultures at the beginning of the historical period from written records. Thus, data from written records may be used to elucidate some elements of descriptive and explanatory models. Consequently models which make use of historical data tend to be more fully structured towards the end of the prehistoric period.

Unfortunately, Philippine ethnographic data are limited for the period of first contacts with the Spanish and for the lengthy period of contact between the Philippine aboriginals, the Chinese, Arabs and Indians. The quality of the sources varies and it is often difficult to know whether they relate to the inhabitants of specific islands of the Philippines or are general descriptions covering the whole of the archipelago.

A number of salient facts emerge from a cursory study of these accounts and they need to be taken into consideration when framing descriptive and interpretative models.

### The End of the Prehistoric period

The Philippines did not emerge from the prehistoric period suddenly. Chinese traders probably started to arrive in the Philippines as early as the 9th or 10th century A.D. (Beyer, 1921: 926; Fox, 1967b: 41) but there is the possibility of much earlier contacts with Indian and Arabian traders as long ago as 500 B.C. (see Jocano, 1975a: 137-145). The nature of these very early contacts is unknown, and it is not until much later, between the 13th and 15th centuries A.D., that more detailed information is available. By this time trade seems to have become more formalized and very extensive (see De Morga, 1609: 177ff.), not only with the Chinese, but with Japanese, Portuguese, Malays, Cambodians and Siamese.

The degree of influence of these early foreign traders on Philippine culture, whether coastal or inland, must have varied considerably throughout the Philippines. There is a need to establish their effects on a regional if not an island by island basis, and in particular, to determine their influence on language, ethnic origins, material culture, social and religious customs. Thus, models will need to take into account the possibility that there have been considerable variations in regional cultures over the last two to three millennia.

Perhaps the most extreme example of cultural conservatism is the Tasaday. In spite of 400 years of colonization by foreign powers, this small group of people managed to remain hidden from the modern world until just recently (Fox, 1976; Nance, 1975).

### Malay Settlement of the Philippines

There are a number of traditions relating to recent settlement of the Philippines. Visayan sources describe how a group of immigrants known as Datus from Borneo landed in the Philippines, purchased the right to settle there, and then subsequently dispersed over the islands, presumably to mix and produce hybrid ethnic groups (Beyer, 1921; Santaren, 1956). The traditions further suggest that well organized political communities already existed in the coastal areas of the Philippines before the arrival of the Malay peoples. However, the ethnic affinities of these pre-Malay groups is not certain though some of the aboriginals were Negrito peoples.

## The Persistence of Cultural Traits

The early records describe a number of cultural traits that have persisted to the present day in several areas of the Philippines. These include:

* the use and manufacture of the bolo (Morga, 1609: 81);
* the use of the umbrella to provide protection from the sun (Beyer, 1921: 928);
* the use of gold to repair and ornament teeth (Quirino and Garcia, 1958: 417);
* the construction of houses made from bamboo on raised platforms (ibid.: 413);
* the manufacture of tuba (Morga, 1609: 77);
* cock fighting (Quirino and Garcia, 1958: 411; Pigafetta, 1521 in Jocano, 1975b: 80);
* rice farming and slash and burn agriculture.

Such traits must have their origins deep in the prehistoric past, certainly beyond the initial period of Spanish contact. The practice of repairing the teeth with gold is one example that can be documented from the archaeological records (Janse, 1944: Plate 12; Legaspi, 1974: 9).

## The Types of Settlement in the Philippines

The earliest Spanish sources describe Negrito peoples living in several of the Philippine islands: Luzon (Morga, 1609: 74; Quirino and Garcia, 1958: 390), the Visayan Islands (Colin, 1663: 44-45), and in Mindanao (Combes, 1667: 125). The remnants of these groups still survive today (e.g. Gloria, 1939; Rahmann, 1955; 1958; 1963; Rahmann and Maceda, 1955; 1962). In general these Negrito peoples seem to have been hunters and gatherers living in the remote and inland areas of the islands. They used bows and arrows and sometimes practised slash and burn agriculture (Chirino, 1604: 217; Colin, 1663: 45; Combes, 1667:125; Morga, 1609:75). A number of sources suggest that the Negrito peoples were "the original inhabitants" (e.g. Chirino, 1604: 217; Colin, 1663: 45; San Antonio, 1738: 302). Other peoples are described living around the sources of rivers (Colin, 1663: 46-47; San Antonio, 1738: 296ff.) and could be distinguished from more sophisticated or civilized peoples living in large villages at the mouths of major rivers along the coastal plains (Chirino, 1604: 219; Morga, 1609: 114-128; San Antonio, 1738: 306). Substantial houses and complex social and political systems were observed in the coastal areas e.g. at Visayan towns (Plasencia, 1589: 175-176; Quirino and Garcia, 1958: 407ff.).

These sources then, give an indication of where major settlements were located during the 16th and 17th centuries and suggest that at least three types of settlement pattern, possibly linked with different ethnic groups, were already established during this period. There were Negritos living in the inland and forested areas, traders and agriculturalists living inland along the rivers and complex agricultural and semi-urbanized stock rearing societies, engaged in widespread commerce and living in areas adjacent to the mouths of rivers along the island coasts.

Communication and Sea Transportation

In considering questions of cultural diversity and change it is essential to know the level of marine technology available to island populations. Thus, a group of people who may have been forest dwellers or slash and burn agriculturalists, on acquiring river or ocean going vessels may have migrated to the coast or to the rivers to develop new economies based on the exploitation of marine and riverine resources, and trade.

Possession of boats would have enabled trade goods and excess produce to be more efficiently transported and traded elsewhere. It should be remembered that in prehistoric times much of the Philippines was covered by dense forests so that the rivers and seaboards were the only convenient and practical means of transport and communication. The presence of foreign merchant seamen in an area may have had similar effects, introducing new technologies and trade goods which sponsored adaptive industries, promoted and established individuals to positions of power and influence, enabled local produce to be moved from one area to another, and fostered political alliances on a limited but effective scale.

Whilst intrusive elements could have filtered into the Philippines overland during the Pleistocene period when sea levels were much lower than they are today, from about 8,000 B.P. onwards sea transportation would have been essential for trade and transmission of ideas throughout the Philippine islands. We have seen that foreign traders probably came to the Philippines during the first millennium B.C. and by the 9th or 10th century A.D. this trade was well established. We also know that Filipinos and foreign traders possessed several kinds of seacraft, including ocean going vessels, at the time of the Spanish Conquest.

Summary

In the context of the foregoing comments it is now possible to summarize broad guidelines for framing models of Philippine prehistory:

* the models should utilize historical and ethnographic facts as far as the evidence will allow, both for filling in details of models and for generating model building hypotheses;
* the models need to be dynamic allowing for regional and cultural diversity at any period in time;
* the models need to make allowance for external influences from different places at different times in prehistory;
* the models need to be regularly structured, and be sufficiently refined to take account of variations in local environment;
* the models need to take account of the limitations of the archaeological evidence, and to ensure that all the archaeological evidence is utilized and not just selective aspects;
* the models need to take cognizance of the geography of the area, and the technological innovations that would have been needed to bring about different forms of cultural and environmental change.

In the long term, a systems model may prove to be the most useful (Bender, 1975: 28ff.; after Clarke, 1968). The Philippine islands can be classified initially into physiographic units (e.g. coastal, plain, uplands, river valleys etc., see Figure 6.2) and similar units from adjacent islands can be grouped (e.g. Eastern Panay and Western Negros, or Eastern Negros and Western Cebu). Each of these arbitrary physiographic units may then be considered as systems interacting with other systems, until such time as evidence is forthcoming to suggest that other groupings might be more useful. The full corpus of archaeological evidence within each system or region can then be assembled, assessed, and interpreted. Finally, data from each subsystem can be compared for each chronological period.

ARCHAEOLOGICAL DATA AND STONE TOOLS

In Appendix 1, we have reviewed the literature describing key Philippine sites that have yielded stone tools for the purpose of defining the stone industries and identifying the changes that have taken place in stone tool technology over the past 30,000 years (Figure 6.1). The stone tool evidence is considered in conjunction with other data to fill the requirements of a simple systems model (see Table 6.1). This review indicates that at the present time there are insufficient archaeological data to enable us to construct a detailed model of any system or region in the Philippines for any period of prehistory.

It can be seen that there are only a dozen or so relevant sites and these are widely spread in time and in space. Not enough sites have been investigated to enable us to build up regional sequences or to define variations in the local stone industries. In several instances only one site type has been excavated in any one region, which is not sufficient to establish the range of local stone technologies either in time or space.

In spite of the limitations of these data a number of observations which are pertinent to model building can be made. There seems to be evidence for at least two basic stone industries in the Philippines.

Smash and Grab Industries

These industries are widespread throughout the Philippines from the Cagayan Valley in the north to Sanga Sanga Island in the Sulu archipelago in the south (Table 6.2). Moreover, in some areas we know that the industries have had a long history; at Tabon Cave from 20-30,000 years B.P. to about 8,500 years B.P., and Musang Cave in northern Luzon 12,000-4,000 B.P., at the Sohoton site on Samar Island from 12,000 B.P. until about 300 B.P., and on Mindanao it appears that a smash and grab technology was still being used by the Tasaday up until a few years ago (Fox, 1976: 4-5). In passing, it is perhaps significant that an essentially similar stone industry has been found even further south at the site of Paso in the vicinity of Lake Tondano, northern Sulawesi, Indonesia, dating from about 6,000 B.C. (Bellwood, 1978a; 1978b). The Philippine industries have the following features:

* extensive use is made of pebbles of crypto-crystalline quartz;
* the flakes are not deliberately shaped;
* hard hammer percussion is normally used to manufacture the flakes, although they may also have been produced by the anvil technique (cores broken on an anvil);

* the absence or rarity of retouch;
* in some assemblages, a preference for the use of conchoidal flakes as tools;
* a tool/waste flake ratio of about 10%;
* a blade production of less than 5% of the assemblages;
* very little, if any core preparation;
* cobble tools are sometimes associated with the flake assemblages.

Typical examples of smash and grab tools are illustrated in Figures 6.4-6.7.

If the Cagayan sites turn out to have a great antiquity then it is possible that the earliest smash and grab industries may be linked with a range of cobble tools, and with the hunting and butchering of large mammals which are now extinct.

It is apparent that there may be considerable variations within the smash and grab industries (refer Tables 6.2 and 6.3). The degree of edge modification or retouch (as opposed to utilization), the level of association of flake artefacts with cobble tools, and the strength of the tendency to choose conchoidal flakes for tool usage, are three features that merit more intensive investigation. In addition, there is a strong possibility that there is variation in the level of flake usage, no doubt depending on the availability of raw materials and the intended functions of the tools. Thus, the Madrigal 1-12 and Pintu sites have relatively high tool/waste flake ratios (1.0 and 0.3 respectively) whilst the Tabon (0.11 - 0.17), Lapuz Lapuz (0.08), Laurente (0.07 - 0.09) and Sohoton (0.11) assemblages have much lower ratios. At Pintu, 9% of the tools are retouched, at Lapuz Lapuz (LL/1), 18%, sites in the Cagayan Valley up to 40%, and at Bungiao none are retouched.

As a further illustration of diversity within this technological tradition, the weight distributions of samples of crypto-crystalline quartz waste flakes from three widely separated sites are shown in Figure 6.3. Although the majority of waste flakes from these assemblages fall in the 0-1 g weight range, in practice there are significant differences in the distributions. For example, the Tabon waste flakes are on average much larger than those in the other assemblages. Similarly, the lengths of crypto-crystalline quartz flake tools from Tabon range between 3.5 and 11.6 cm and 80% are greater than 5 cm long (Fox 1970: 37). At LL/1, the average tool length is only 3.3 cm and 60% of the tools lie in the range 2.7 - 4.2 cm.

Interesting features of these industries are their associations with other cultural traits (refer Table 6.1). Clearly they survived the introduction of agriculture, animal husbandry, metal working and pottery making. In the later sequences, the stone industries are associated with earthenware pottery, and possibly with jar burials (e.g. at the Bungiao site). In all places where they appear, the industries seem to be linked with hunting and gathering, though associations with agricultural/horticultural societies cannot be ruled out. Some of the most recent smash and grab sites (such as LL/1) must have been contemporary with more elaborate archaeological assemblages and with more

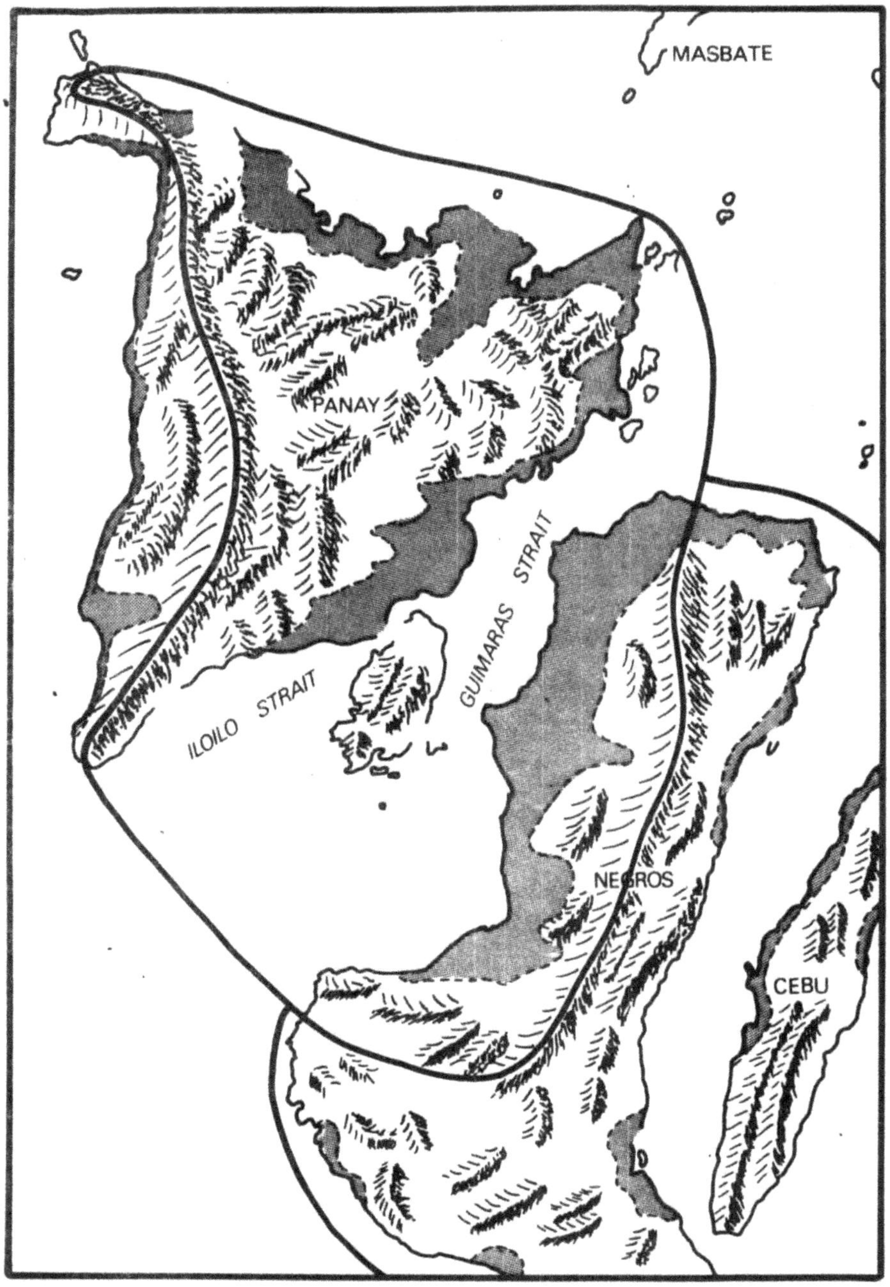


KEY :

- Regional Boundaries (based on physiographic criteria)
- Costal lowlands and river valleys
- Mountanous areas and rolling foothills

Figure 6.2 Systems analysis: an example of a physiographic province. The Visayan islands can be divided into analytical units or physiographic provinces which might be used for systems analysis.

TABLE 6.2 Select Characteristics of some Philippine stone industries

| Site | Type of Industry | Blades | Retouching of Flake Tools | Presence of Cobble Tools | Evidence of Silica Gloss | Evidence of Organic Staining | Evidence of Core Preparation | Presence of Preformed Flakes | Tendency to make Tools from Conchoidal Flakes | Tendency to make Tools from Waterworn Pebbles |
|---|---|---|---|---|---|---|---|---|---|---|
| LL/1 | Smash and Grab | Rare ( 3%) | 18% | Present but rare | - | - | Present but rare | - | Strong (59%) | Strong |
| Tabon | Smash and Grab | Fairly rare ( 6%) | Rare | Present but rare | ? | ? | - | - | Strong (59%) | Strong |
| Guri | Blade ? | Present | Common | ? | ? | ? | ? | ? | ? | Strong |
| Duyong | Blade | Common | - | | ? | ? | Present | ? | ? | Strong |
| Madrigal-12 | Smash and Grab | Rare | Common ( 40%) | ? | ? | - | - | - | ? | Strong |
| Musang | Smash and Grab | Rare | Rare | ? | Present | ? | ? | ? | ? | ? |
| Sohoton | Smash and Grab | Rare | Rare | ? | Present but rare | ? | - | - | Strong (35%) | Strong |
| Barrio-Y-Barra | Blade | Common | Common | ? | Present | Common | Present | - | ? | Strong |
| Sanga Sanga | Smash and Grab | Rare | Rare | Present | ? | ? | - | - | ? | Strong |
| Bungiao | Smash and Grab | Rare | Rare | ? | ? | ? | ? | - | ? | ? |
| Cabatuan Iliolo | Smash and Grab | Rare | Rare | Present | ? | ? | - | - | ? | Strong |
| Pintu | Smash and Grab | Rare | Rare ( 9%) | Present but rare | ? | - | - | - | ? | Strong |
| Laurente | Smash and Grab | Rare | Rare | Present but rare | Present | ? | - | - | - | Strong |

TABLE 6.3 Some Characteristics of "Smash and Grab" Industries from select Philippine Archaeological Sites: Crypto-Crystalline Quartz Artefacts only

FLAKE TOOLS:

| Site | Layer No. | No. of Flake Tools (T) | CONCHOIDAL FLAKES: No. of Conchoidal Flakes | CONCHOIDAL FLAKES: No. with Cortex | CONCHOIDAL FLAKES: No. with Hinges | CONCHOIDAL FLAKES: No. of Blades | BLOCK FRACTURED FLAKES: No. with Cortex | BLOCK FRACTURED FLAKES: No. with Hinges | BLOCK FRACTURED FLAKES: No. of Block Fractured Flakes |
|---|---|---|---|---|---|---|---|---|---|
| LL/1 | 3 | 96 | 57 | 13 | 4 | 3 | 7 | 2 | 39 |
| | (%) | | (59) | (14) | (4) | (3) | (7) | (2) | (41) |
| Tabon | B | 4 | 1 | 2 | 1 | 1 | 1 | 0 | 3 |
| (Sample) | (%) | | (25) | (50) | (25) | (25) | (25) | (0) | (75) |
| | C | 28 | 20 | 5 | 4 | 2 | 1 | 0 | 8 |
| | | | (71) | (18) | (14) | (7) | (4) | (0) | (29) |
| Laurente | 1 | 17 | 2 | 1 | NR | 0 | 3 | NR | 15 |
| (Sample) | (%) | | (12) | (6) | | (0) | (17) | | (88) |
| | 2 | 12 | 1 | 0 | 0 | 0 | 3 | 1 | 11 |
| | (%) | | (8) | (0) | (0) | (0) | (25) | (8) | (92) |

WASTE FLAKES:

| Site | Layer No. | No. of Flake Tools (W) | CONCHOIDAL FLAKES: No. of Conchoidal Flakes | CONCHOIDAL FLAKES: No. with Cortex | CONCHOIDAL FLAKES: No. with Hinges | CONCHOIDAL FLAKES: No. with Blades | BLOCK FRACTURED FLAKES: No. with Cortex | BLOCK FRACTURED FLAKES: No. with Hinges | BLOCK FRACTURED FLAKES: No. of Block Fractured Flakes | T/W x 100 |
|---|---|---|---|---|---|---|---|---|---|---|
| LL/1 | 3 | 1233 | 288 | 113 | 84 | 23 | 528 | 103 | 945 | |
| | (%) | | (24) | (9) | (7) | (2) | (48) | (8) | (77) | (8) |
| Tabon | B | 38 | 13 | 2 | 3 | 0 | 6 | 0 | 25 | (11) |
| (Sample) | (%) | | (34) | (5) | (8) | (0) | (16) | (0) | (66) | |
| | C | 168 | 68 | 15 | 27 | 10 | 21 | 6 | 100 | |
| | (%) | | (40) | (9) | (16) | (6) | (13) | (4) | (60) | (17) |
| Laurente | 1 | 181 | 35 | 5 | NR | 4 | 52 | NR | 146 | |
| (Sample) | (%) | | (19) | (3) | | (2) | (29) | | (81) | (9) |
| | 2 | 175 | 35 | 6 | 12 | 2 | 32 | 2 | 145 | |
| | (%) | | (20) | (3) | (7) | (1) | (18) | (1) | (83) | (7) |

NR - Not Recorded

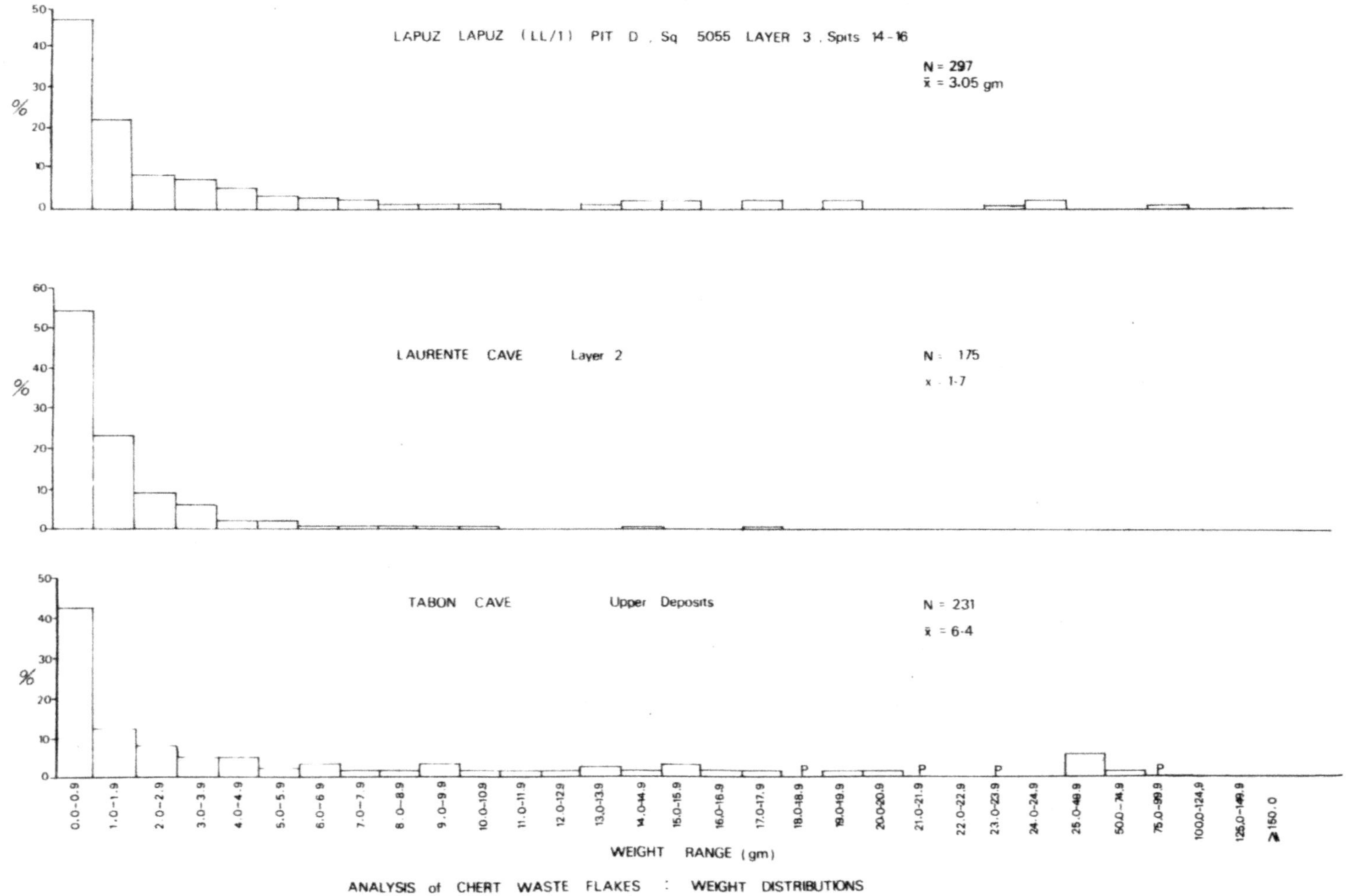


Figure 6.3 Smash and grab industries from three Philippines sites. Weight distribution of waste flakes.

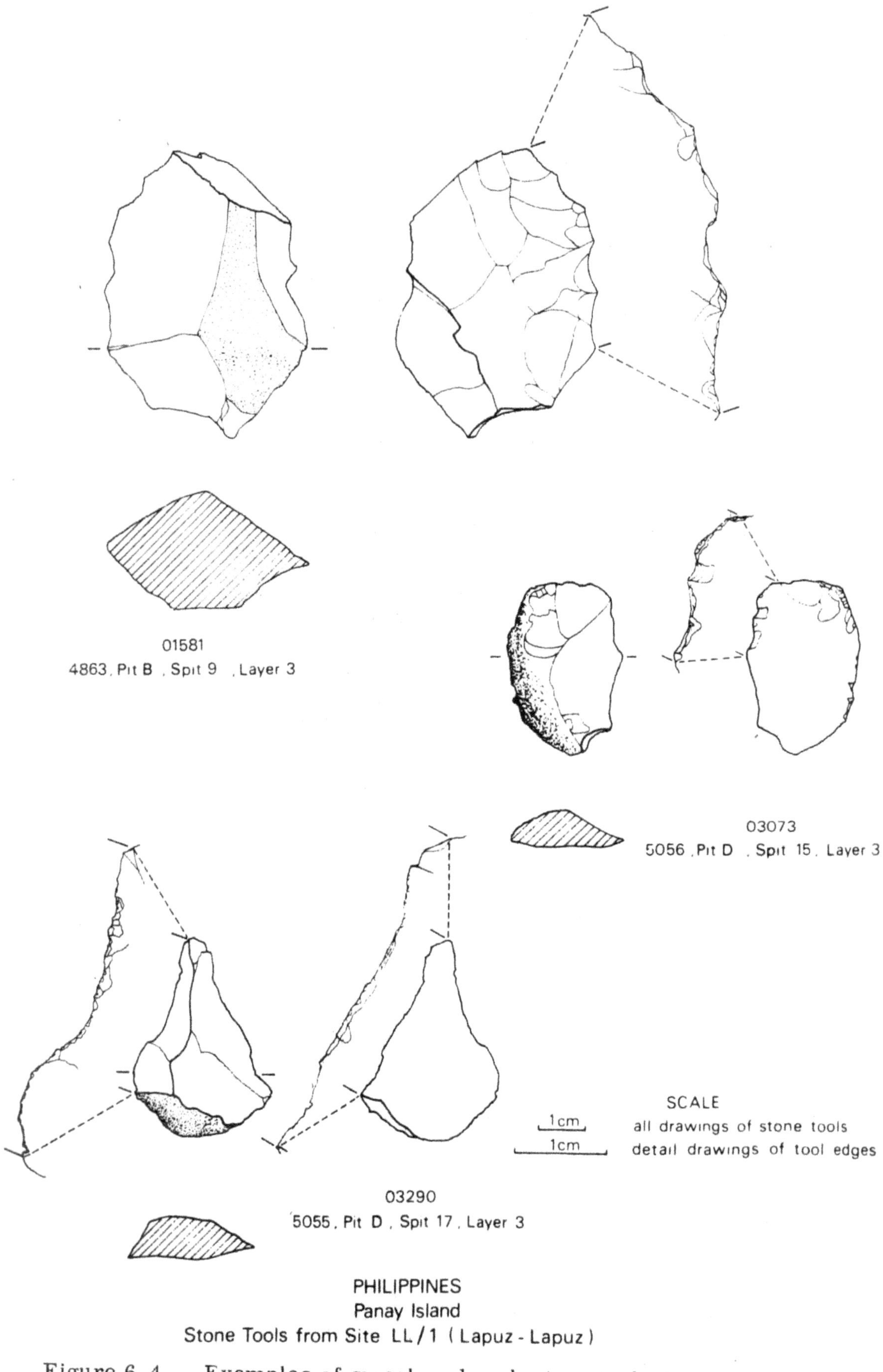


Figure 6.4 Examples of smash and grab stone tools from Lapuz Lapuz, Panay Island.

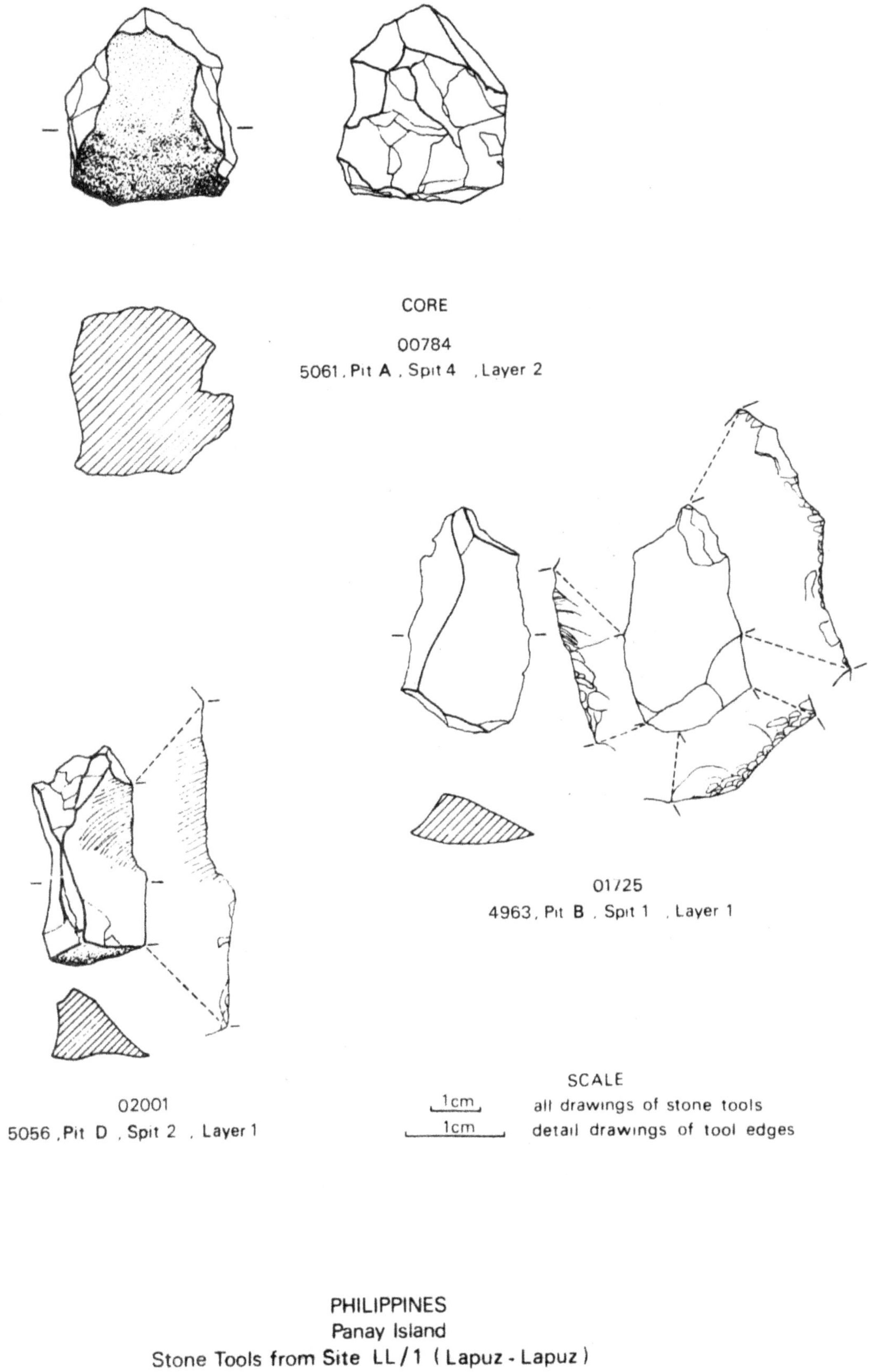


Figure 6.5 Examples of smash and grab stone tools from Lapuz Lapuz, Panay Island.

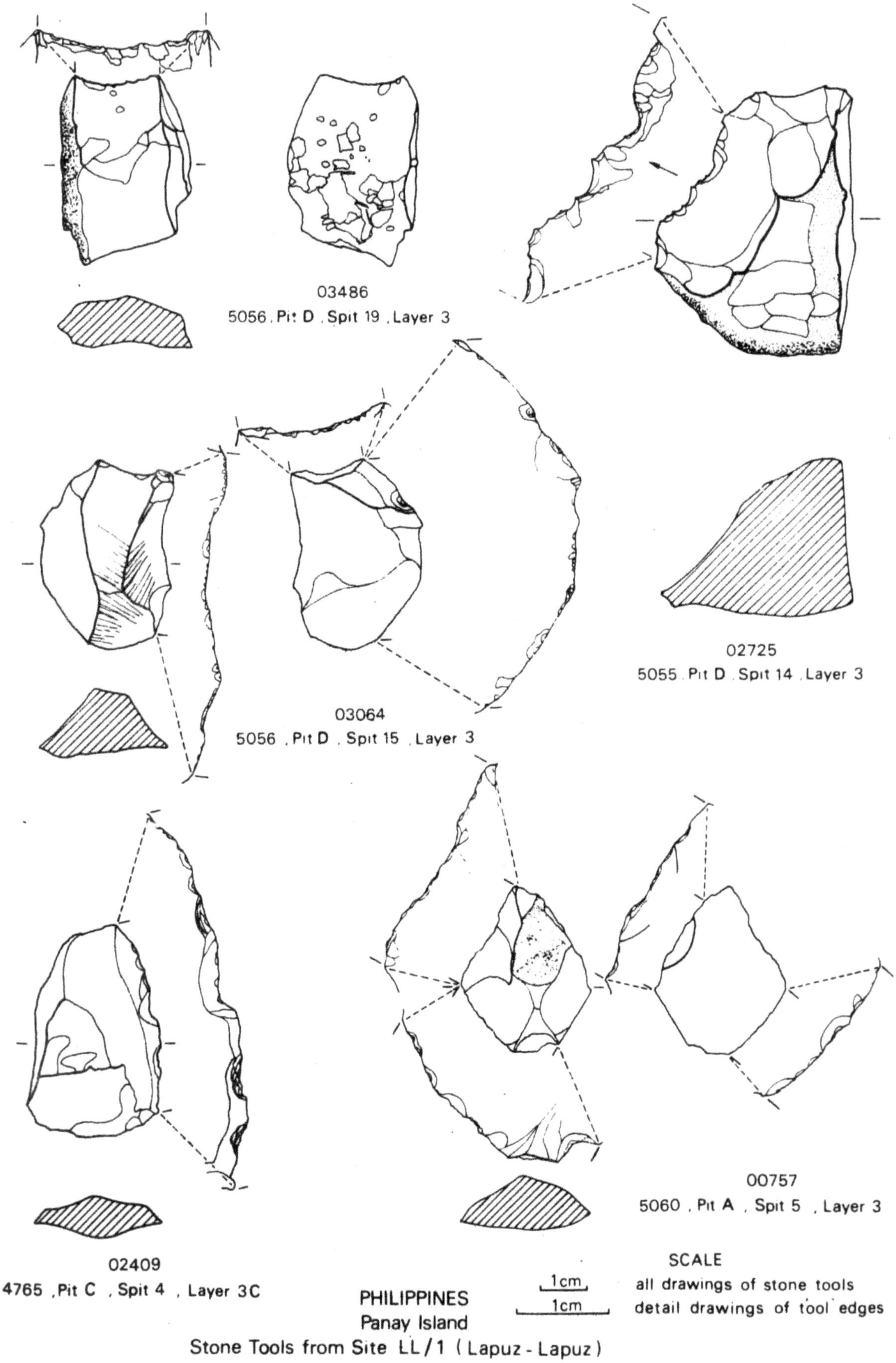


Figure 6.6 Examples of smash and grab stone tools from Lapuz Lapuz, Panay Island.

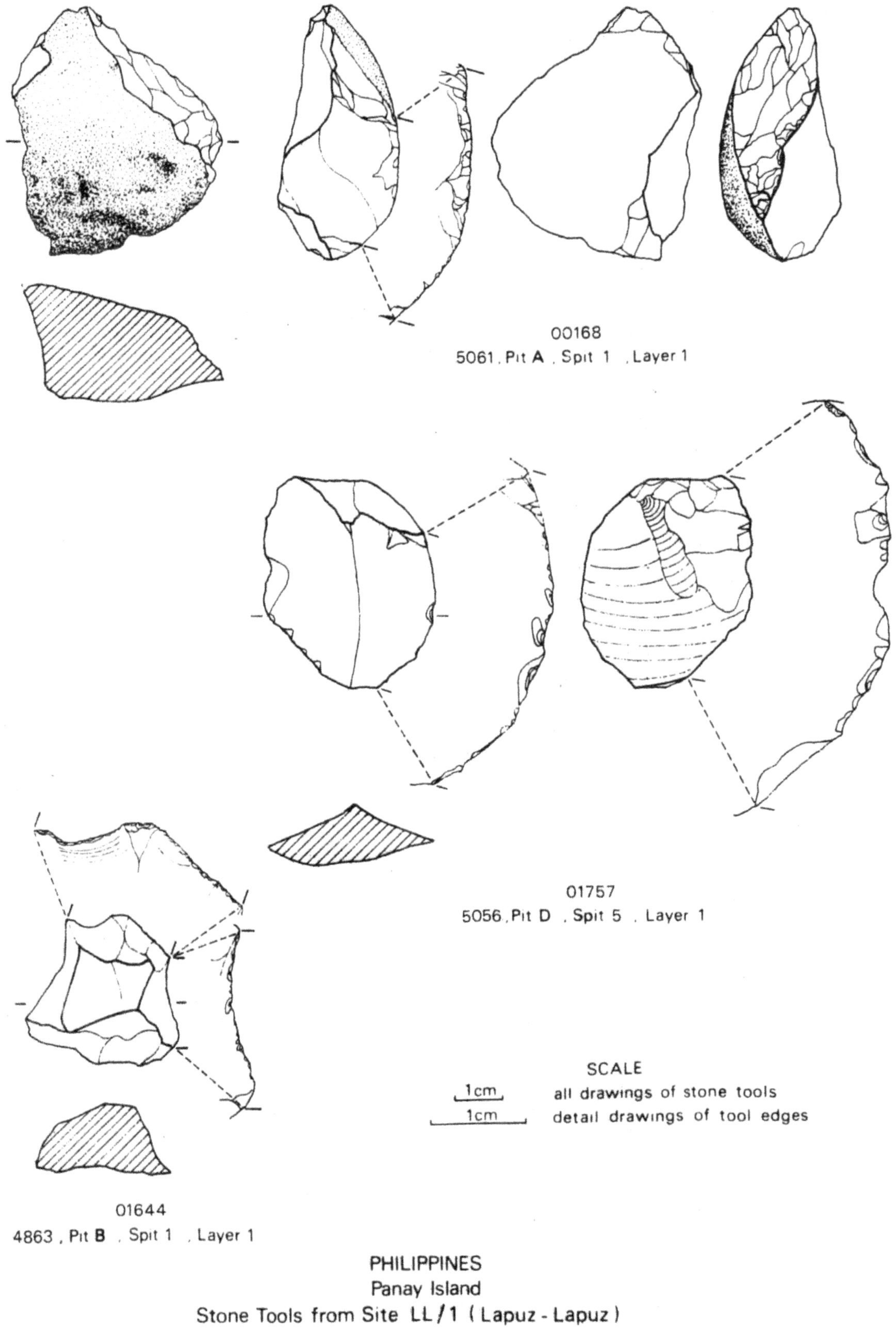


Figure 6.7 Examples of smash and grab stone tools from Lapuz Lapuz, Panay Island.

sophisticated cultures. However, it is not clear at the moment whether the industries are seasonal/functional manifestations of more complex societies or belong to a remnant contemporary population of aboriginals who had escaped the machinations of civilization.

Blade Industries

Several blade industries have been identified at Philippine sites and in general, they seem to appear much later in the archaeological records (Tables 6.1-2). However, like the smash and grab industries they are found over a wide area from Palawan in the west, to southern Luzon and Buad Island in the east. With one exception, the blade industries have never been properly described. A common feature seems to be the presence of prepared cores (Table 6.2). It is clear that blade manufacturing techniques did not evolve from, nor necessarily replace, block fracturing technology. After blade technology was adopted in the Philippines, the two technologies co-existed for a lengthy period of time. For example, the stone industry at Duyong cave on Palawan seems to be an extension of the smash and grab industry at Tabon. The significant difference is the presence of smaller flakes and some bladelets, but like the LL/1 and Tabon smash and grab industries there is little emphasis on retouch in the Duyong assemblages. Likewise, although the more recent assemblage at Guri cave is thought to contain some blades, it is essentially a smash and grab industry with a lot more retouch than is found on the Tabon and LL/1 artefacts. At Barrio-Y-Barra on Buad Island, however, there is a much greater emphasis on both the final production of blades and on edge modification with retouch.

Again the functional associations of the blade industries are not clear, although hunting and gathering is one strong possibility. On present evidence the technology appears to have its origins in the period before the introduction of agriculture and other sophisticated technologies (e.g. pottery making) but later in the prehistoric sequence blade tools are contemporary with, if not directly associated with, jar burials, trade ceramics, metal, edge-ground tools and shell adzes (e.g. Duyong Cave on Palawan). It is interesting that shell adzes have been found in association with a much earlier smash and grab industry on Sanga Sanga Island. Again blade industries have been recorded in sites dating between 1,400 and 1,600 A.D., but to date they have not been described. Consequently, at this time it is not possible to say anything about their variability in the Philippines.

The links between blade and smash and grab technologies are hazy. It is possible that they were used interchangeably by aboriginal populations, depending on the intended functions of the tools. Alternatively, they may be linked with two different groups of peoples. Another possibility is that the smash and grab tools were used by hunting and gathering peoples and the blade tools by slash and burn agriculturalists (cf. Bellwood, 1978a: 263; 1978b: 12-13). On Buad Island there are indications on the blades that they were used for processing plants.

## CONCLUSIONS

In the foregoing we have attempted a selective review of Philippine prehistory and archaeology. In so doing a number of problem areas have been identified. We have seen that current models used for organizing and interpreting Philippine prehistory need to be substantially modified to take account of the realities of Philippine geography and archaeological, ethnographic and historical evidence. The need to develop dynamic, regionally based models has been strongly advocated, and concurrent with this approach we have suggested that new data gathering strategies are required if these types of models are to become a reality.

In the context of model building the archaeological evidence relating to flake and blade industries has been reviewed in the context of an assumption that they may be used to monitor some aspects of culture change. The review has demonstrated that there are insufficient data to enable regional models of Philippine prehistory to be constructed at present. However at least two basic stone technologies can be identified and these have been described, though the relationships between these industries remain obscure (Table 6.1). This study has demonstrated that these industries have a widespread geographical distribution and has shown the variability of the smash and grab facies. Such variability is an argument for regional model building and highlights the dangers of generalizing the archaeological data over wide areas of the Philippines.

If Philippine prehistory is to be rationalized in the future, there is an urgent need for full and comprehensive site reports to be published. It would be a good beginning if the sites reviewed in this paper were published in full.

In the process of model building there is no reason why traditional views of Philippine prehistory as gleaned from the ethnographic and historical sources need be ignored or abandoned. Indeed, these data may be used with great advantage for generating research questions in regions where detailed fieldwork is to be conducted.

We will conclude this paper with an example of a model which has been constructed from historic and ethnographic data, and which can be tested when archaeological data are available from a systematic regional survey. The historical evidence hints that the first inhabitants of the Philippines were Negritos who were predominantly hunters and gatherers. It might be assumed that early in prehistoric times hunters and gatherers inhabited the richest food resource zones such as the coastal areas and/or inland river valleys which were frequented by large grazing animals (now mostly extinct). At some stage, perhaps concurrent with the adoption of the edge ground axe, hunting and gathering may have been supplemented by slash and burn agriculture, possibly beginning along the coastal margins. However, in time the slash and burn technique spread inland along the major river systems. Throughout prehistory and from time to time, the individual islands must have been visited by traders, explorers and colonizers. Communication between islands and individual bands of hunters and gatherers varied, and was dependent in large measure on the level of maritime and riverine technology at the time. After the close of the Pleistocene period, the first contacts with outside peoples must have been in coastal areas probably near the mouths of rivers where there were convenient anchorages for boats. It is near such localities that we might expect to find evidence for the growth of permanent settlements served by agricultural communities. It is in the coastal areas also that we would anticipate more complex political and economic institutions to evolve, since it was the communities in these areas that controlled the trade networks. Expansion and annexation of hunting grounds belonging to the adjacent indigenous populations would follow, possibly leading to colonization and development of the interiors of the islands. As this happened some adaptations would have been necessary since the colonists, who were adjusted to coastal and riverine economies, would need to learn how to exploit the resources of forest and inland environments. Trade networks incorporating the inland areas would evolve with the rivers playing an increasingly important role in communications. As the process of urbanization continued, political institutions became more complex and in the coastal areas the cultural inventory expanded at the expense of artefacts which may have been in common usage previously. Thus we would anticipate that bronze and iron tools would replace those made of stone.

Historical models, such as the one described here, can be superimposed on a regional, systems model; the latter can be used for ordering the archaeological data and testing hypotheses, the former may be used for generating hypotheses. Other models can be generated by referring to comparable situations outside the Philippines (see Bender, 1975). What we need now are the data to test these models.

## APPENDIX

## A REVIEW OF THE ARCHAEOLOGICAL EVIDENCE FROM SELECT PHILIPPINE SITES

### PANAY ISLAND: EXCAVATIONS AND SURFACE COLLECTIONS

#### LL/1 and NA/I Stratigraphy and Chronology

Since 1976 the authors of this paper have been surveying sites and conducting excavations in Iloilo and Capiz Provinces of Panay Island (Coutts and Wesson, 1978a; 1978b). Two relevant sites have been excavated—the first is a cave site known as Lapuz Lapuz (abbreviated LL/1) situated in the Bulabog-Putian National Park, near Dingle, Iloilo Province. The second is a rock shelter called Naulan (abbreviated NA/1) near Dumalag in Capiz Provvince.

LL/1 has two major layers and its deposits consist of shell middens, mixed with the remains of animals, earthenware sherds and considerable numbers of stone artefacts. Only one fragment of metal was found during the excavation and this may be intrusive. Two radiocarbon dates are available from layers 2 and 3 of Lapuz Lapuz: 880±200 B.P. (SUA-991) and 1010± 90 B.P. (SUA-990). The main difference between the two layers is that the lower levels contain less pottery.

NA/1 was test pitted. Six individual layers were distinguished and like LL/1 it is a very concentrated shell midden about a metre deep and contains large numbers of animal bones, plus a few fragments of pottery and a stone flake assemblage. No metal or Chinese ceramics were recovered from this site. One radiocarbon assay has been obtained from a shell sample, but it is far too old, given the artefacts in the deposit. Since the NA/1 assemblage, in most respects, is essentially the same as that of LL/1 it will not be discussed further.

#### Material Culture

Stone Tools

The stone industry may be termed a block fractured or smash and grab industry. Pebbles from a nearby river were brought to the site and then roughly smashed into block pieces. Most of the stone flakes were manufactured from cryptocrystalline quartz, which occurs in a variety of colours, and is sometimes banded. Flakes struck from tektites were also recovered. The stone tools were manufactured at the site, since cores, waste flakes and tiny shatter flakes were all recovered in situ. The industry has the following features (refer Table 6.3):

* Retouch is rare, and use wear, common;
* Tools are not shaped; there is some evidence for platforming of cores, but generally speaking this is rare;
* Blades are rare in the assemblage (3% of the tools are blades, and 2% of the waste flakes are blades);
* The tool makers tended to choose conchoidal flakes for tools;
* Most of the tools were probably used for the production and maintenance of implements associated with hunting and gathering;
* So-called cobble tools were present in the assemblage.

The weight distribution of waste flakes in major layers is shown in Figure 6.3. It typifies the distribution expected for a block fractured industry, with a very high percentage of flakes falling in the lower weight ranges. In more controlled industries worked from cryptocrystalline quartz the distributions tend to be more flattened.

Bone and Antler Tools

Several ground and polished bone and antler tools were recovered. Some of these are probably awls. The utilization is consistent with the presence of deer remains in the midden.

Pottery

Approximately 1200 earthenware sherds were recovered from LL/1. There were no imported porcelains. The majority of the sherds derive from pots and bowls which were used as cooking vessels and are undecorated. However, the few decorated pieces have incised decoration or impressed rims and some were red slipped. The pottery has been moulded and finished with paddle and anvil. There is evidence that a bound paddle and anvil was used in some instances. The vessels have been smoothed, polished or slipped. The vessels have been fired at a low temperature, probably less than 500° centigrade. Given the nature of this evidence (which includes only 11 decorated sherds), it is not possible at this stage to determine the cultural affinities of this pottery.

<u>Economic Evidence</u>

The inhabitants of LL/1 collected large numbers of shellfish, particularly the terrestrial species <u>Rhysota rhea</u> and the freshwater <u>Thiara (Melanoides) tuberculata</u>. The latter must have been carried to the site from the local river which is distant about 5 km. The skeletal remains mixed in the midden are often burnt and have butchering marks, which suggests that most of these materials have been introduced by man. The dominant species represented include pig, deer, bat and rat. However the remains of monkey, snake, lizard, frog, turtle and crab are also present. Although the site was carefully sifted for plant remains, no evidence was found.

At this stage, the economic evidence at the site suggests that the inhabitants were hunters and gatherers. The site was probably a transitory staging camp for hunting expeditions by an agricultural people living on the plains below.

The Cabatuan area, Iloilo Province

There are several surface assemblages from this area collected by officers of the Philippine National Museum. These assemblages consist predominantly of cryptocrystalline quartz flakes which were manufactured by the smash and grab technique.

PALAWAN ISLAND EXCAVATIONS

Fox (1970: 167; Table XIII) has located 84 cave sites on Palawan of which 29 are situated at Lipuun Point, and 35 in Quezon and 19 in the El Nido areas respectively. Of these sites only three might be described as habitation sites: Tabon and Guri (Chamber A) Caves at Lipuun Point (ibid.: 9, Table 1) and Duyong Cave located about 11 km north of Lipuun Point (ibid.: 53). They are associated with stone tools and have been sampled by excavation.

The only published descriptions of the methods used to excavate these sites are offered in a footnote (ibid.: 7, footnote 2). The sites were dug according to natural and cultural levels and the "most rigorous archaeological control" was used. The deposit was screened through 1/8th inch (3.2 mm) mesh sieves, which means that while a fair percentage of the fine debitage must have been collected, a significant percentage may have been lost.

Fox has used a chrono-technological model to date his sites—a combination of radiocarbon dates and the presence/absence of type fossils, materials and technological traits. Sites have been seriated according to the following criteria: origin of pottery; type of vessel; material and origin of ornaments; material and origin of tools; type of technology; presence of coffin burials; presence of cord/paddle marked pottery and the presence of blades.

Tabon Cave: Stratigraphy and Chronology

Of the three "habitation" sites, the Tabon Cave is by far the most important, as Fox contends that the stone assemblages from this site date beyond 30,000 B.P. to about 8,000 B.P. (Fox, 1970: Table 2; 1967a: 99, Table 1). He distinguishes a number of horizons which include 5 flake assemblages numbering I-V. From these horizons he has excavated 2,500 artefacts (Fox, 1967a: 103). There are a number of anomalies in the published reports which make interpretation of these data difficult. The stratigraphic relationships at Tabon Cave are summarized by Fox (1970: Fig. 6, 8-10), but it is difficult to relate plans to sections and vice versa. Nor is it easy to understand the layer allocations. In his Fig. 9 (1970) for example, why is there no layer allocation for the "hard laminate floor"? How far does flake assemblage II extend horizontally into layer 1, and are the layer boundaries as precise as the sections indicate? Again, referring to his Fig. 10, it seems that flake horizon III appears in layers IV, V and VI. Finally, it seems rash to declare layer V a habitation floor or that it constitutes a flake assemblage when only three flakes have been excavated from it (ibid.: 28).

In order that the Tabon assemblages can be compared directly with others elsewhere in the Philippines, the authors have analysed samples of chert artefacts from layers B and C, although the precise relationship between these layers and those cited in the published text has not been established. The results are summarized in Table 6.3 and illustrated in Figure 6.4. Particular attention has been focused on the analysis of the waste flakes.

## Material Culture

### Stone Tools

Even though there are a number of stratigraphic anomalies associated with the various flake assemblages (Fox, 1970: 25, 27), it would appear that the flake components of assemblages I to IV are essentially similar. They have the following characteristics (Fox, 1970: 31-33):

* the flakes were struck from waterworn pebbles of essentially chert or cryptocrystalline quartz;
* there was no conscious attempt to shape the tools;
* the technique of direct, hard hammer percussion was used though the anvil method seems to have been employed as well;
* tools have little or no retouch;
* there is no evidence of core preparation;
* tools are highly irregular, though there are some round ended scrapers that seem to recur. Tools are essentially waste flakes that have been selected from a heap of debitage (refer Table 6.3);
* our analysis of the Tabon industry indicates that blades are present but scarce (Table 6.3). Nine percent of the tools are made on blades, and 5% of the waste flakes are blades. Fox (1970: 33) calls the flake tradition Tabonian.

### Metrical Analysis and Flake Core Classification

Fox (1970) has classified 337 chert artefacts excavated from a single layer (flake assemblage III) from a volume of approximately 6.3 $m^3$ (calculated from his figures) giving a flake density of about 53 flakes per cubic metre. He summarizes his flake-core classification in tabular form (1970:35, Table 4) distinguishing between primary flakes which are potential tools (flakes which appear to be similar morphologically to flakes that have been used as tools), flake tools (those with some evidence of utilization), flake tools with retouch, unused lumps of chert (presumably waterworn pebbles brought to the site), and waste flakes (debitage from tool making activities). The lengths of a sample of 56 tools from flake assemblage III were used and these varied between 3.6 cm and 11.6 cm with 80% being more than 5 cm long.

Some comments on Fox's analysis are in order. First of all the figures given in his Table 4 are misleading. Elsewhere (Fox, 1970: 38) he relates that hundreds of tiny chips of chert were recovered when sieving with a fine wire mesh and that these were "not counted in the statistical analysis". He does not say what criteria were used to divide the smaller from the larger flakes. Thus any attempt to calculate the tool/waste flake ratio from his figures (which is about 0.29 in this case), is meaningless. Indeed a calculation of the ratio following a re-analysis of a sample of the Tabon material (Table 6.3) yielded ratios of 0.11 and 0.17 respectively. These figures are much closer to the anticipated values for this type of assemblage. Moreover the division of the waste flakes into two major categories—primary flakes (or potential tools), and waste flakes, is of questionable value. The reasons given barely justify the division.

Fox expresses concern that there are so few unaltered waterworn pebbles in the assemblages (1970: 35) that he is led to conclude that many of the cores must have undergone decortication at the core collecting site before they were brought to Tabon Cave. Given the mode of flake production and the types of materials being used, the small number of unaltered cores is predictable. A large number of waste flakes are frequently produced when this type of core is struck, since the material tends to fracture in an uncontrolled manner. Thus core materials are exhausted quickly and can be discarded. Flakes can be selected from the debitage for use as tools. This problem can be examined further by counting the percentage number of waste flakes and tools with cortex, since decortication of cores elsewhere would presumably remove much of the cortex . Of 205 waste flakes re-analysed from the Tabon assemblages, 21% had some cortex (Table 6.3). This is much lower than for layer 3 of LL/1 (51%) or layer 1 at Laurente (31%) but it is still significantly high.

Pottery Complex

At Tabon, sherds were recovered from within layers 1-3, (refer Fox, 1970: Fig. 9-10) and the excavations at Tabon and Duyong yielded at least "1,500 whole or reconstructed earthenware vessels and tens of thousands of sherds" (ibid.: 75). At no stage is it made clear how the pottery relates to the stone tools. However, it would appear from the sections (refer Fox, 1970: Fig. 10) that there is pottery associated with stone tools and possibly stone assemblage 1, although the date for the latter (9,250 ± 250 B.P.) would make it the earliest pottery in the Philippines. It appears that the sherds derive from earthenware jars used for burials and grave goods. At any rate, in spite of the stratigraphic anomalies, the most recent phase of cave usage at Tabon is younger than 9,250 B.P. and is associated with a jar burial complex.

Fox's (1970: 105ff.) study of the Tabon jar burial complex concentrates on six sites in the Tabon area, excluding the Tabon Cave. The earliest jar burial site dates from Fox's "late neolithic" or from about 710-890 B.C. at Manunggul A (ibid.: 111). None of these sites had stratified sequences (ibid.: 68) nor has Fox found any cave on Palawan with stratified jar burials. Consequently there is a gap between the latest date in the Tabon Cave and the earliest dates for jar burials in the Tabon area and this discontinuity in the Tabon sequence needs to be explained.

Associated with the jar burial complex from Tabon Cave were four cornelian beads, 42 jade beads (Fox, 1970: Table XII), jasper beads (ibid.: Table X), a round wooden object believed to be a jar cover (ibid.: 73), bronze fragments and glass beads (ibid.: 124) which from comparative evidence suggests that the cave was used from Fox's Neolithic to the Early Metal Age (ibid.: 162-163).

There are no descriptions of the pottery from the Tabon Cave though there is one reference (ibid.:87) to examples of organic glazed pottery from the site. Perhaps some of the nine pottery types (ibid.:78) he describes are present in the assemblage and if so, some may have been utilitarian. The stratigraphic positions of the pottery and the associated items (such as beads) are not given. However it can be assumed the Tabon pottery has the general characteristics of Fox's Tabon Pottery Complex. Briefly, this pottery is sand tempered, hand moulded, formed with paddle and anvil and scraped and smoothed before firing (Fox, 1970: 77-78). Most of the pottery is plain, but has been decorated and modified (see Table VII), using the techniques of polishing, impressing (use of cord marking and carved paddle), incising (rare), painting (rare) and organic glazing (rare). Fox claims that the Tabon pottery tradition is more closely related to traditions in South East Asia, particularly Niah, Borneo and the Sa-Huynh of Indo-China, than to the Kalanay Complex in the Philippines (Fox, 1967a: 92; 1970: 98). Fox found that some of the key features of Kalanay Complex pottery such as the presence of incised decoration, the absence or rarity of paddle impression, and some distinctive forms, were absent or rare in the Tabon assemblages. Consequently, he suggested that the Tabon pottery is ancestral to the Kalanay, which he assigns to his Developed Metal Age (Fox, 1970: 100-103).

Fox's arguments need more qualification. He has been able to establish that a distinctive pottery tradition exists on Palawan and that there are local variations. Basically, its distinctive features are a matter of decorative technique (impressing with cord and/or paddle), the absence of one or two distinctive pottery types (massive footed vessels and tripod and tetrapod forms —Fox, 1970: 103), which occur in other complexes, and the variety and form of the decorative motifs. Other factors such as vessel form, method of manufacture (slipping, polishing, use of paddle and anvils etc.) are traits shared in common. Another important consideration concerns the justification of comparing pottery complexes on the basis of decorative elements which are geographically so widely separated.

In making such comparisons Fox has tended to assume that the stimuli for making pottery (including the surface treatment used and the method of manufacture) would probably have come from somewhere outside the study area (Fox, 1970: 94, 161-162) and secondly that his vessels are representative of the full range of earthenware in use throughout his late Neolithic to the Developed Metal Age. There is already accumulating evidence that the earlier Filipinos were more familiar with sophisticated technologies than originally thought (e.g. copper and iron smelting—see Hutterer, 1973: 35; 60; Tenazas, 1973).

The degree of foreign influence on pottery making is arguable, and there is no firm archaeological basis for Fox's contention that many exotic items, including bronze, iron, jade, pottery making, ear pendants, beads and a highly developed culture of the dead (See Fox, 1970: 94-95, 98, 122, 123, 134, 139, 162-166) were brought into the Philippines by small groups of people from time to time. There are alternative explanations, some quite complex, but like Fox's hypotheses, they cannot be tested because of the meagre nature of the archaeological evidence available.

It is important to bear in mind that only three of the 84 caves examined were "in any way habitation sites"; that the sherds from Tabon (one of the three) were from or associated with a burial complex and that the rest of the 84 caves are burial sites. Indeed the pottery percentages presented (Fox, 1970: 79; Table VII) are from four locations, only one of which is a habitation site (Duyong).

Thus the pottery from these sites may not be typical and truly representative. Indeed since it is associated with funerary rites it is probably atypical. Again, as we have pointed out above, the Kalanay, Bau and Novaliches complexes have been largely defined on the basis of surface collections and excavated sherds from disturbed archaeological deposits. Whilst most of this is also funerary material it would seem wise at this stage to refrain from making inter-regional comparisons until such time as regional sequences have been established from better documented data.

Another way of looking at this problem is to consider the representation of excavated vessels over time. Using the data presented by Fox, (1970: Table VII), his samples are distributed as follows:

| Period | Dates | No. of Vessels | Vessels/ 100 Years |
|---|---|---|---|
| Late Neolithic | 1,500-500 B.C. | 78 | 7 |
| Early Metal Age | 500-200 B.C. | 200 | 66 |
| Developed Metal Age | 200 B.C.-900 A.D. | 205 | 21 |

Clearly, the sample is small. Fox is obviously assuming (without valid reason) that local pottery traditions were conservative and changed slowly with time. Thus, if we were to assume that a local population of say 500 persons was in the habit of burying their dead in local caves, and if there was some cultural continuity from 1,500 B.C. until 900 A.D. and if only one person died per year, there would be approximately 2,400 burial jars nnd associated offerings in the caves (even allowing for multiple burials in jars). This figure of course would be an absolute minimum. Although there were considerable numbers of jars in the caves we have an impression from Fox's reports that most of the sherds derived from non-jars, and that jars in the Lipuun area can be counted in their hundreds and not thousands. Consequently we might anticipate that most burials took place in areas other than caves. Certainly, Fox (1970:64), on the basis of the lack of archaeological evidence from cave sites, argues that Neolithic man did not use cave sites for habitation and preferred to live in open sites.

In summary, we would anticipate that stone tools from the Tabon Cave will be associated with jar burials and earthenware pottery at the most recent end of the Tabon sequence—unless we accept that people representing two different cultures utilized the same cave. We may also assume that the stone tool makers were familiar with pottery making and the range of pottery and technology which Fox has defined as his Tabon Pottery Complex.

Economic Evidence

The Tabon excavations have yielded considerable quantities of faunal data (Fox, 1970: Fig. 9-10), but no detailed reports are available. Nevertheless it would appear that at the time flake assemblage I was being deposited (about 9000 years B.P.) the sea was still some distance from Tabon Cave, as no marine shells have been found in the upper levels of the excavation.

Flake assemblages, I, II and III are associated with large numbers of animal bones, mainly of bats and birds, and freshwater shells, plus a few bones of larger animals including pig and deer (Fox, 1970: 25-26, 28, 38). The question of whether or not these remains were introduced into the cave by man is not discussed. The absence of the remains of pig and deer from most levels invites explanation, since these animals must have been relatively common in the area during prehistoric times. Indeed if one were to exclude the "few bones of large animals", the other faunal remains (small animals) are those which one might expect to find in caves uninhabited by man. No evidence of plant foods was found at Tabon (Fox, 1970: 39). Until such time as the faunal data are properly reported, little more can be said about the economic evidence.

Guri Cave - Chamber A: Stratigraphy and Chronology

Fox (1970: 45ff., Fig. 13) offers no details of Guri Cave other than a floor plan. The maximum depth of the cultural deposit was approximately 70 cm (Fox, 1967a: 108) and a radiocarbon date of 4,070±80 B.P. was obtained for marine shell at a depth of 60 cm below the floor of the shelter. The terminal date for occupation of the site is not known. These deposits appear to be much more recent than the uppermost levels of the Tabon Cave (about 8,500 B.P.) which would thus overlap in time with Guri.

Material Culture

Stone Tools

Chamber A of Guri Cave contained a rich assemblage of flake tools (Fox, 1970: 48; 1967a: 107-108) which seems to be a variation on the Tabonian. Fox claims that the main distinction between the two assemblages is a reduction in the size of the flake implements, an increase in the number of tools with secondary retouch and the presence of large blade tools in the uppermost levels. Although no cores appear to have been recovered from Guri, a reduction in the size of the waste flakes would be consistent with more controlled and deliberate attempts to fashion tools. Thus, if a significant number of blades were produced, Fox may be right in his assertion that a new flake technology associated with the preparation of cores was present. However small numbers of blades and blade-like flakes can also be manufactured by the smash and grab technique (see Table 6.3) so that the Guri assemblages need to be thoroughly assessed before the industry is characterized.

Pottery

Fox (1970: 47) does not say whether the deposit in Chamber A is associated with artefacts other than those made from stone. However he does suggest (ibid.: 45, 51) that sherds were found in both chambers (presumably A and B) though it is not clear whether they are sub-surface and/or surface finds. A jar burial complex was investigated in Chamber B and said to be much later in date (between 2,250 and 2,450 years B.P.—Fox, 1970: 47). It falls within Fox's Tabon pottery tradition.

Economic Evidence

Like Tabon Cave, many bone fragments were recovered from these excavations, but they remain to be described and analysed (Fox, 1970: 45, 47, 50).

According to Fox a greater variety of animals, including pig and deer, are represented at Guri. His explanation, that the inhabitants of Guri possessed better weapons for hunting and trapping, may be correct, but no evidence is presented to support this claim. Other explanations are equally valid: inhabitants of Guri may have hunted a different range of animals from a different range of habitats, or seasonal and social factors may have influenced the types of tools used. Marine shells were also found at Guri.

Comments

In many ways this site is potentially one of the most fascinating and important in the Lipuun area, yet the details available are so fragmentary that one hesitates to use the data. It would be helpful to know whether the upper levels of the deposits in Chamber A have pottery and to obtain a radiocarbon date for the upper layers. Another radiocarbon date for the lower levels of the site would be appropriate also as it is not good practice to build a chronology around a single radiocarbon date.

Duyong Cave: Stratigraphy and Chronology

More detail is offered for this important site (Fox, 1970: 53-65; Figs. 16 and 17). Four cultural layers have been distinguished, the lower two being associated with a "small flake and blade" industry. Three radiocarbon dates are available for the site, two from Neolithic levels (4,630±150 B.P. for a burial and 5,680±80 B.P. for a habitation level) and one for a flake and blade assemblage (7,000 ± 250 B.P.). Flake and blade assemblages are associated with the two lowest layers (Fox, 1970: Fig. 17) and it is not clear from which of these levels the older date comes. The most recent layers are linked with a jar burial assemblage.

The presence on jar burial layers of domestic artefacts such as spoons (Fox, 1970: 140, 147, 153), cooking pots, "bronze slag" (ibid.: 140), moulds for making adzes (ibid.: 134), bivalves with use-sheen (ibid.: 119) could suggest that at times the site was used for habitation as well as for burial. However, such objects also tend to be characteristic of jar burial furniture (cf. Bellwood, 1978a: 269). Unfortunately it is very difficult to identify activity areas in jar burial caves because there is usually no stratigraphy (Fox, 1970: 68). Many of the caves have been looted, both recently and in prehistoric times, and there is no way of gauging how much evidence has been lost (ibid.: 72, 130; 147). However, Fox (1970: 73 ) claims that remains of domestic utensils and weapons were not common, which must suggest that the Cave was probably a specific activity site rather than a habitation area.

Material Culture

Stone Tools

The stone industries associated with the lowest two levels consist of large numbers of flakes and blades made from chert (Fox, 1970: 59). Tools have not been retouched, though there is evidence of core preparation and blade production. Fox does not explain why he calls the Duyong assemblage "small flake and blade", as this description would seem to cover the Tabon and Guri assemblages as well. It is not clear whether the Neolithic layers of the site (including the one associated with jar burials) are linked with flakes and blades.

A feature of this industry is the production of unretouched blades from prepared cores. This is unusual, as blade production is normally associated with retouch. However, a similar blade industry has been excavated by Bellwood in the Talaud Islands of Indonesia (Bellwood, 1978a; 1978b), though other blade industries found in Indonesia are associated with retouch (Glover, 1977). Perhaps detailed analysis of the Duyong stone artefacts will reveal more numerous flakes and blades, or prepared cores. In either case, the assemblage would begin to taken on a more Tabonian character.

Other Artefacts

No further data is available about the associations of artefacts in the lower levels. However there is considerably more information from the upper levels. Fox's (1970: 60) Early Neolithic burial (4,630±250 B.P.) is linked with such items as Tridacna shell adzes, shell pendants, and a polished stone adze. A possible neolithic habitation floor (5,680±80 B.P.) was associated with "hearth-like areas", a Tridacna gouge, and a number of shell discs (ibid.: 62).

The jar burial layers (dated to the Early Metal Age) were extremely rich, yielding the remains of over 200 vessels which were associated with the remains of stone adzes, bronze artefacts and iron fragments (ibid.: Table VI), "Ling Ling-O" ear pendants, bracelets and beads made of gold, jade and stone (Fox, 1970: 119, 126).

Economic Evidence

The flake and blade tools were associated with a midden which contained a variety of shellfish that come from a range of environments (forest, mangrove and marine). Few bone fragments were recovered from the midden. These included the remains of pig and hawk, crab and stingray, but no deer (ibid.: 56). However there is no evidence that these remains were introduced by man.

Comments

Like the Guri excavations, insufficient data are published from the Duyong Cave to enable it to be properly evaluated. What can be said is that an early flake and blade industry existed at Duyong, but it is not clear whether it should be regarded as an industry in its own right, an industry ancestral to that of Guri, or a variation of Fox's Tabonian. Certainly it is premature to associate the Duyong flake assemblage with a hafting technology and the use of composite tools (Fox, 1970: 48, 50). Fox's argument that core preparation necessarily leads to blade production, hafting and composite tools, is not sound given the evidence he presents. How blade tools were used is a matter for determination for each particular culture, and it seems unwise to invoke ethnographic and archaeological parallels from outside the Philippines at this stage. Again Fox is perhaps premature in suggesting that two distinct peoples are linked with the Tabonian and Duyong stone industries. The two sites seem to overlap in time and they are geographically close to one another. Indeed, the foraging zone of prehistoric hunter gatherers residing at Tabon would probably have included the Duyong Cave. There seem to be some differences in the economic patterns evident at the two sites—at Tabon the remains of small mammals are common, at Duyong there are few vertebrate remains and lots of shellfish. At both sites there is abundant evidence for stone tool manufacture, but we do not know for what purpose they were using the stone tools. One possibility is that the two industries are variations of a single stone tool tradition where the variations result from different ranges of activities at the two sites.

## LUZON: ARCHAEOLOGICAL SITES IN THE CAGAYAN VALLEY, NORTHERN LUZON

### Cabalwanian Sites: Preamble

Having found a site on Palawan Island with an antiquity of more than 30,000 B.P., taking Philippine prehistory well back into the Pleistocene, it was providential that even older archaeological evidence seemed to be present in the Cagayan Valley in northern Luzon. Fox and his associates commenced work in the area in 1971 and since that time several preliminary reports have appeared describing results of on-going research (Fox *et al.*, 1972; Fox and Peralta, 1974; Fox, 1973). Many of the conclusions reached in these reports, although tentative, are highly provocative and for this reason they have merited close scrutiny (see Coutts *et al.*, 1976). Such for example is the claim that *Homo erectus* stalked and killed a megafauna which roamed the Cagayan Valley some hundreds of thousands of years ago.

Although excavated data are available from several sites in the Espinosa, Madrigal and Wanawan areas of the Cagayan Valley, their interpretation is made difficult because of stratigraphic anomalies at the sites (see Coutts *et al.*, 1976).

### Pebble Tool Assemblages

On the basis of stone materials collected from a number of excavations and on a large number of surface collections, Fox and Peralta (1974) have demonstrated that pebble and cobble tools are associated with flake industries. Further, they propose that the stone tool tradition, which they call Cabalwanian, persisted for the duration of the middle Pleistocene and then emerged as the late Pleistocene Tabonian, retaining much of the character of the earlier Cabalwanian. The implications are that these traditions key with the range of variation found elsewhere in South East Asia and that they are a parochial aspect of Philippine stone technology. In line with their re-assessment of the evidence in northern Luzon, Fox and Peralta (1974) have proposed a new classification for the pebble tools found in that area. The scheme is derived from a study of surface collections and is held to be preferable to the methods advocated by Movius (1948) and Sieveking (1958) of classifying pebble tools. There is no doubt that a more meaningful classification of palaeolithic pebble tools is needed (after Fox and Peralta, 1974: 110); however, it is questionable whether that proposed by Fox *et al.* (1972) is an improvement on those currently in use. It seems most unrealistic when a new classification of eight types is proposed on the basis of 63 unprovenanced artefacts (*ibid.*: 112, Table II), and in spite of a stated aversion to speculative functional classifications, the authors have not resisted this temptation.

Their classification of eight types (A-H) is based on manufacturing criteria, attributes such as unifacial/bifacial flaking, shape of the working edge, tool cross-section, and so on. Fox and Peralta (1974: Appendix I) provide some data for the cobble tools. Of these there are 37 tools from the Espinosa and Madrigal sites which appear to be sufficiently well provenanced to justify further analysis (see Coutts et al., 1976). Using the data for these tools it was discovered that by plotting width/length ratio against weight. Fox's types A and F tended to form a group with width/length ratios of approximately one. Type C tends to be oblong, as do most of the other types, including some of the Type A specimens. The sizes of all the specimens vary.

Fox's types can be considered to be within a range of what would ordinarily be called unifacial hand axes but otherwise poorly differentiated and highly variable tool types (types A, C and F). These unifacial hand axes show continuous variation towards specialized extremes (see Figure 6.8). Thus, type B would usually be called a pick, and type E a chopper. Type D, in addition to being rounded on one side and worked on both pointed and lateral sides, from Fox's illustration, is a "sumatralith". Type G is supposed to be a bifacially flaked core, but there are no measurements given. However, judging from the illustrations, the type G "bifacial pebble/cobble tools with alternate flaking", would appear to be part of a continuum terminating in a "flake-core tool" category.

In summary, it would seem that the assemblages consist mainly of a generalized unifacial hand axe tradition with some rare and specialized forms as follows:

| | | |
|---|---|---|
| A, C, F | (unifacial hand axes) | 30 |
| B | (picks) | 4 |
| E | (choppers) | 2 |
| D | ("sumatraliths") | 2 |
| G | (bifacial core tools) | 2 |

It should be re-emphasized that only data from the Espinosa and Madrigal sites have been used here. It is also assumed that specimens with accession numbers prefixed by 71-G are all from the Espinosa site (including one printed as 71-7) and that the Madrigal site is 71-M. With some reservation, five specimens which were patinated or rolled or both, have been included in this analysis.

As Fox et al. (1972) have pointed out, terms like "chopper" or "sumatralith" leave a lot to be desired taxonomically, but an argument for their continued use is that we have some idea of what is meant by such terms. The problem is basic: there is little point in trying to devise a new taxonomy for Hoabinhian artefacts until the significance of what is being classified and measured is established.

The significance of the Cagayan Valley finds is perhaps best seen in the context of comparable assemblages. Some decades ago Tindale (1937) and McCarthy (1938; 1941) recognized the similarities of some tools in early Australian assemblages with those in South East Asia, and, at a later date, this line of enquiry has been pursued by Matthews (1966) and McBryde (1976). There are, for example, overall similarities in respect of size for Hoabinhian type assemblages. Figures 6.9 and 6.10 show the range of the width and length ratios respectively for 37 implements from the Madrigal and Espinosa sites. They seem to fit into the spectrum of Hoabinhian implements.

Flake Tool Assemblages

The Cabalwanian flake tool assemblages are said to have the following features:

* they are a mixture of flake and cobble tools;
* there is no evidence of core preparation;
* there is no evidence that flakes were deliberately trimmed to particular shapes and forms;
* there are a few "types" and the latter tend to be "high domed, flake-core scrapers with flat bases, also high angled and step trimmed around the circumference of the working edge" (Fox and Peralta, 1974: 123);
* the ratio of utilized/retouched tools to waste flakes tends to be high (of the order of 1 for the Madrigal 12 site—(Fox and Peralta, 1974: Table IV);
* the working edge of the tools tend to be low angled;
* a high percentage of the flake tools have retouch (40% in the case of the Madrigal 12 site);
* the tools appear to have been manufactured from waterworn pebbles.

Clearly this industry shares common characteristics with the Tabonian, but in other respects it differs as, for example, in the high percentage of retouched flakes, the presence of cobble tools, and the very high tool to waste flake ratios. If the tools were made locally, then a much lower tool/waste flake ratio could be anticipated, providing of course that the small waste flakes had not been dispersed in antiquity and that the sieving techniques used during the excavation were sufficiently elaborate to recover the smallest flakes.

The metrical description of the Cagayan Valley flake tool assemblage is confusing. Fox and Peralta (1974, Appendix II) present a table of results for the analysis of flake materials from the Madrigal 12 site. Measurements include length, width and thickness for some 50 utilized and retouched flakes. The measurements are listed in millimetres, and if this is correct, then about half the flakes would be under 4 mm long. However, an examination of illustrations of representative flake tools suggest that these measurements should be in centimeters, so that assemblages are not classified as microlithic.

If measurements are in centimetres, then the length distribution (Fig. 6.11) peaks between 2 and 7 cm. However, the distribution is very ambiguous, since the sample number is too small to give a reliable result. Again, the large size of the flake tools is striking. The sieving methods used at the excavations are not described, but if only coarse sieves were used, smaller tools would not have been captured, and this would have the effect of increasing the average size of the excavated tools.

When length is plotted against weight there are two clusters (Figure 6.12). The first is an elliptically shaped distribution of small flakes. In addition, there seems to be another group of larger flakes which are heavy in proportion

| | POINTED-END-STRUCK | | STRAIGHT-SIDE-STRUCK |
|---|---|---|---|
| Flat both sides | Type A | Type F | Type E |
| Flat/rounded each side | Type C | Types D, G | |
| Rounded both sides | Type B | | |
| G S | | | |

NOTE: The dashed line separates the generalised from the specialised.

Figure 6.8 Reorganization of Fox's types of cobble tools from the Cagayan Valley.

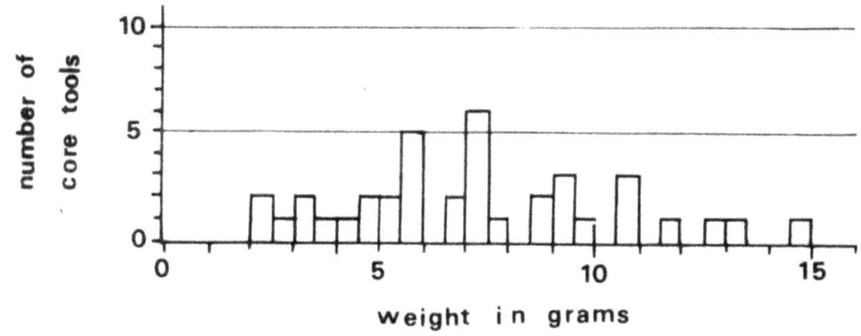


Figure 6.9 Analysis of cobble tools from the Cagayan Valley sites: weight frequency distribution (N=37)

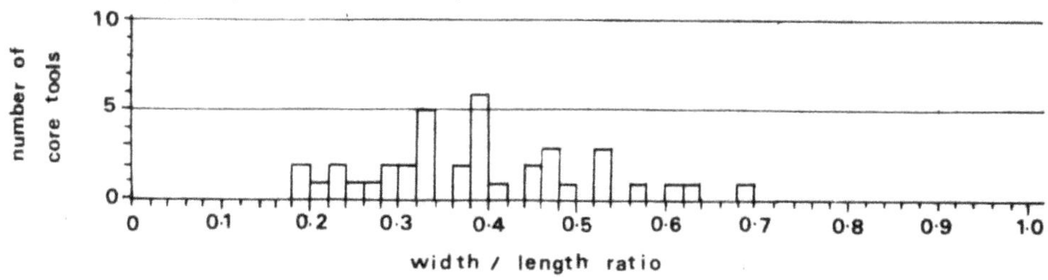


Analysis of cobble tools from the Cagayan Valley sites: width/length ratio. N = 37

Figure 6.10 Analysis of cobble tools from the Cagayan Valley sites: width/length ratio (N=37).

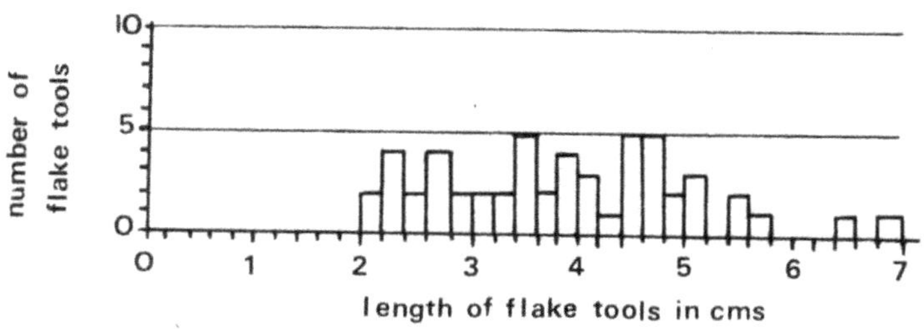


Figure 6.11 Analysis of flake tools from the Cagayan Valley sites: length frequency distribution.

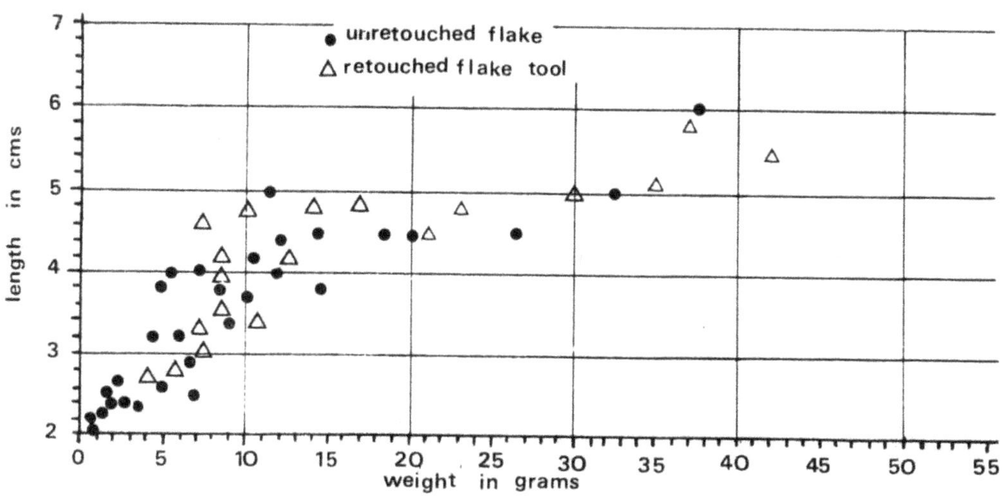


Figure 6.12 Analysis of flake tools from the Cagayan Valley sites: weight plotted against length.

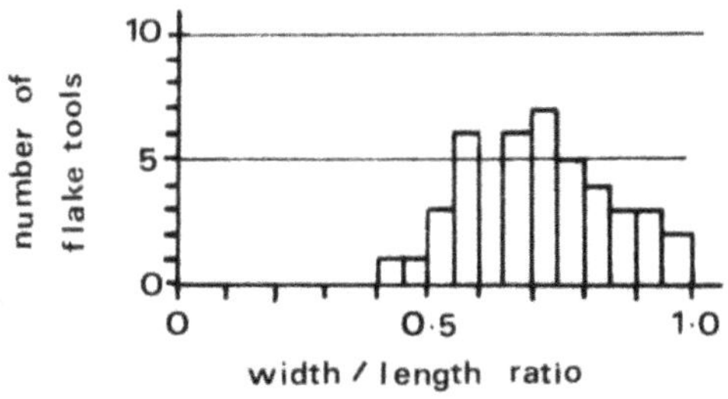


Figure 6.13 Analysis of flake tools from the Cagayan Valley sites: width/length ratio.

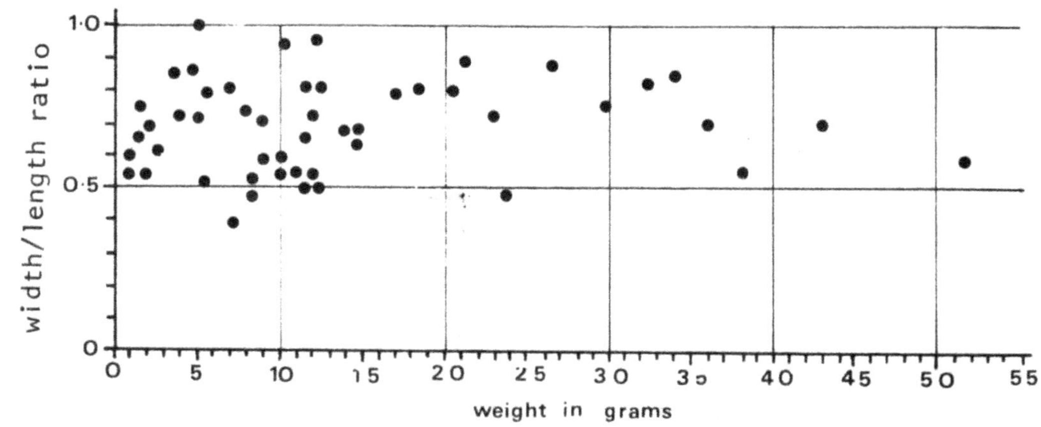


Figure 6.14 Analysis of flake tools from the Cagayan Valley sites: weight plotted against the width/length ratio.

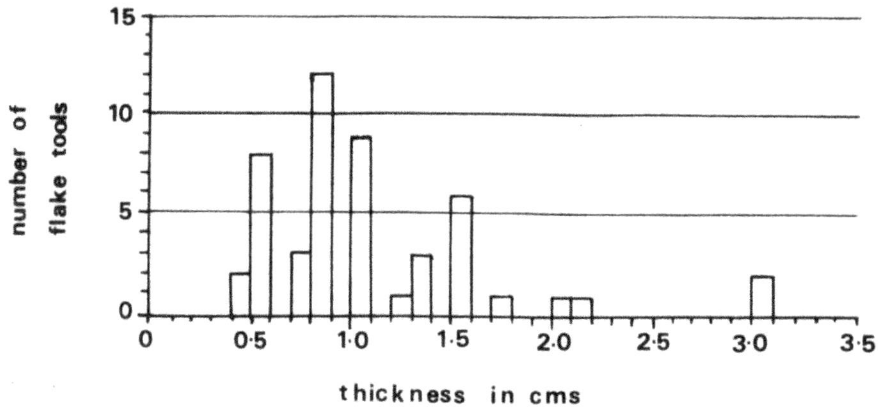


Figure 6.15 Analysis of flake tools from the Cagayan Valley sites: thickness frequency distribution.

to their length. Hence two different facies of a stone industry may be presented and mixed together though the resulting distribution could be attributed also to the selective usage of different materials (say from flaking fine grained and/or coarse grained materials). Certainly, it would be useful to have the materials identified.

Observations of the width/length ratios indicate that nearly all of the flakes can be bracketed in the 0.5-1.0 range (Figure 6.13) with a peak between 0.6 and 0.7. Hence most of the flakes have widths which measure approximately two thirds of their lengths and represent relatively broad flakes.

When the width/length ratio is plotted against weight there is no predominant group of flakes (Figure 6.14). Thus, in this assemblage, the main difference seems to be between thick and thin flakes. A bar graph of thickness (Figure 6.15) suggests there are two groups of flakes; one in the range of 0.5-1.0 cm, and another with a peak of about 1.5 cm. Consequently, it appears that two sorts of flakes were produced: thin flakes, possibly from fine grained cores, and thick flakes, possibly from coarse grained cores. Without more data, however, this line of investigation cannot be pursued further.

### Economic Evidence

It is claimed that stone tools are associated with the remains of extinct animals which were hunted by prehistoric man, but until the stratigraphy of the area has been properly established this interpretation will continue to be controversial (see Coutts *et al.*, 1976).

### Pintu Shelter: Stratigraphy and Chronology

A rock shelter located on the Ngilinian River in the Cagayan Valley has been excavated by Peterson (1974a). It yielded a flake-core industry in the Cabalwanian tradition associated with faunal remains and pottery. The site dated between 2,260±150 B.P. and 3,880±140 B.P. No detailed data are available at present.

### Material Culture

#### Stone Tools

Flake assemblages were recovered consisting of waste flakes, flake and core tools, and cores made variously from quartz, basalt, andesite and more rarely from cryptocrystalline materials. Features of the industry include:

* flakes manufactured by direct percussion;
* no evidence of core preparation;
* retouch is rare; of 866 flake tools only 9% had retouch;
* a high percentage of the flakes were utilized (approximately 30%). However no information is given about the waste flakes and the method of recovery;
* the tools were made locally (from local materials), since there were numerous hammer stones and cores at the site;

* core tools are present, and can be classified into types which are "closely reminiscent of Hoabinhian core types" (Peterson, 1974a: 28);
* Peterson claims that some of the flake tools can be further classified into functional categories such as awls, gravers, burins, scrapers, etc.

Pottery

Some fragments of nondescript earthenware were excavated in the middle and upper levels of the site. The original vessels were coil and ring built and then finished with paddle and anvil. Most of the sherds were not decorated and the vessels are classified on the basis of firing temperature and temper. Only one vessel form was identified.

Bone Awls

Six bone tools were recovered from the site but their function is not known, although Peterson suggests that they may have been used for sewing fibrous materials.

Economic Evidence

Faunal remains were scarce and in bad condition. The remains of deer, pig, turtle, shellfish and rodent were recovered although no evidence is presented to show that they were introduced into the site by man.

Other Sites in the Cagayan area

Over the past few years, investigations have been conducted in several rock shelters in the Cagayan Valley particularly in the Penablanca area near Bagabag. The most interesting aspect of these sites is that they have yielded assemblages similar to those found at open sites in the Cagayan Valley. No reports from these sites have been published.

At Rabel Cave, Ronquillo (1977) has recovered andesite cobble tools, waste flakes, cores, sherds and animal remains. Flakes derive from numerous materials including cryptocrystalline quartz and basalt, although the former tend to be rare. The stone industry is similar to that described by Peterson for the Pintu Cave. It is associated with plain earthenware pottery. Animal bones were present in small numbers and indicate a hunting economy (Ronquillo, pers. comm.).

The Laurente Cave, excavated by Henson (pers. comm.) has yielded stone assemblages which are related to those at Pintu and Rabel. Aspects of the flake industry are illustrated in Figure 6.3 and summarized in Table 6.3. The assemblages of Rabel and Laurente caves are expected to date from the late Pleistocene to the late Holocene.

Near Bagabag, Thiel (n.d.) has excavated deposits in Musang Cave, the deposits dating from about 12,000 B.P. The lower levels do not contain pottery and the upper levels yield sherds dating from about 4-5,000 B.P. Stone flake tools and animal remains are present throughout the deposit. Although there are no published descriptions of the industries, they fall within the typical smash and grab tradition. The people who inhabited the site hunted animals such as pigs, and gathered terrestrial shellfish.

Comments

Much of the vitally important data for the Cagayan Valley is not yet published, and the data published are generally cursory and difficult to assess. Nevertheless, it would seem that there is an increasing body of evidence to suggest that a single stone tool technology existed in this area from the upper Palaeolithic to at least 2,000 B.P. It is in the smash and grab tradition and is associated with the use of modified cobbles. In all cases the cobbles seem to have been gathered from old stream beds or from gravel deposits. Source materials vary, but basically they comprise andesite, basalt and cryptocrystalline quartz.

Whereas the Madrigal and Espinosa sites would seem to be Palaeolithic sites, Peterson has shown that implements indistinguishable from Habinhian have dates of less than 4,000 B.P. in the Cagayan Valley, and are found in association with ceramics. Howvver, Matthews (1966), using large samples and sophisticated statistical methods of analysis, was able to show that differences could be identified between the Australian and South East Asian pebble tool assemblages. At this stage the numbers of core tools from the Cagayan Valley sites are insufficient to test the data with these methods.

At present there is no clear indication of the variability of the stone tools and the manner in which they were used throughout time. There is very little provenanced economic and artefactual evidence for the Upper Palaeolithic, and the data from Holocene sites varies in quality, and is mostly unpublished. However, these data suggest that hunting and gathering were major preoccupations, and it may be assumed that at least some of the stone tools were associated with hunting and gathering activities (cf. Peterson, 1974a; Thiel, n.d.).

During the Holocene, pottery is associated with the stone tool industries, and although new subsistence activities may have accompanied its introduction, the same types of stone tools continued to be made, implying continuity in this aspect of material culture.

The research potential of archaeological sites in the Cagayan Valley is great. To date, comparisons of the Cabalwanian type assemblages from different areas have not taken into account important deterministic factors such as environment and geography which are important in any consideration of problems such as intra-site variability, control and exploitation of local natural resources (following Matthews, 1966: 21). It is clear that the Cagayan Valley sites are especially numerous and produce tools of such abundance that an extensive field programme should produce samples adequate to enable a range of variables to be taken into account.

Samar Island: Sohoton Cave: Stratigraphy and Chronology

Excavation of Sohoton 1 on Samar Island has yielded archaeological materials including flake assemblages which date from 10,500 ± 160 B.P. to 385 ± 105 B.P. (Tuggle and Hutterer, 1972: 11). The stone industry continues unchanged to the uppermost levels, where it is mixed with fragments of earthenware pottery (ibid.: 59, 69). The faunal aspects of these excavations have not yet been reported.

Material Culture

Stone Tools

The industry comprises an assemblage of cryptocrystalline quartz flakes and fragments made in the smash and grab tradition; there are no prepared cores or specialized flake shapes and the flakes were made by percussion from waterworn pebbles. All told, a total of 7,500 flakes were excavated from eight natural layers using a 1/8" sieve mesh. However, to date only 5000 flakes have been analysed, and of these only 110 have been subjected to detailed analysis. The results of the analysis are summarized as follows:

* 7% of the assemblage are tools;
* A significant number of tools (35%) are conchoidal flakes;
* The tools generally have evidence of use wear on one edge and retouch is rare;
* The tools can be divided into two groups—those with low edge angles (average 35°) associated with rounded worn edges, and those with high edge angles (average 56°) associated with minute flaking;
* One flake had evidence of silica gloss;
* 3% of the assemblage consisted of cores, 86% waste flakes and 26% of the flakes were conchoidal.

Pottery

Some 1163 fragments of earthenware pottery were recovered from about 31 vessels. The basic pot shape was a globular form (ibid.: 62 ff.). The vessels were finished with a paddle and anvil, slipped and polished—and the rims seem to have been made separately and joined to the body of the pots. Only one pot had decoration.

Barrio Y-Barra, Buad Island: Stratigraphy and Chronology

This is an open site, and excavation yielded artefacts from three stratigraphic horizons (Cherry, 1978). Altogether 139 tools were recovered. The stone tools appear to be mutually exclusive of the upper levels which contained sherds. The excavation data were complemented by an extensive reconnaissance and collection of surface assemblages which yielded 5,000 stone artefacts. The method of collection is not specified in the report. Test pits were sunk in two separate areas, and the stratigraphy of the two pits was correlated and "confirmed through soil analysis" (Cherry, 1978: 11). However no documentation is presented to support this conclusion.

Cherry claims Barrio Y-Barra is not a habitation site (ibid.: 9, 22) but at no stage does he give criteria for identifying such sites, nor does he suggest a function for the site. However, given that there is evidence of flakes and workshop activity, the site must have at least been a transient habitation site. Stone tools are the only aspect of the site described in detail. One radiocarbon date is available for the upper levels of the site, but it is comparatively recent and Cherry suggests that it does not date the cultural materials (ibid.: 36).

Material Culture

Stone Tools

Ninety nine percent of the assemblage consists of cryptocrystalline quartz flakes (Cherry, 1978: 22) struck from waterworn pebbles. Cherry concludes that he is dealing with a blade industry which he calls Buadian (ibid.: 23, 28). Retouch has been applied by both pressure and percussion. Like the Sohoton, LL/1 and Tabonian assemblages, the tools do not have specific shapes or forms but some cores show evidence of preparation, and a significant number of the flakes (20%) and blades (11%) have organic staining suggesting that many of the tools were hafted. Again, about 10% of the blades had a gloss on their working margins, which implies that they were used for working organic materials. The tools included: heavily retouched scrapers and points; tools with notches, spurs and tangs; borers and tools with burin scars (ibid.: 32-33).

Unlike many of the archaeological reports discussed above, considerable metrical data are presented in this report. However, whilst the basic approach is sound, some of the underlying assumptions behind the analysis are questionable.

Only 139 stone artefacts were excavated compared with greater than 5,000 artefacts gathered from surface sites. On the basis of an examination of the excavated materials and the fact that the excavated flake industry seemed mutually exclusive of the stone adzes and sherds found in the upper levels, he concluded that the flakes gathered from the surface sites must have eroded from the lower deposits which were of Buadian origin. This he saw as sufficient justification for amalgamating surface and excavated artefacts, giving him a sample of more than 5,000 items for analysis.

This procedure is questionable. Firstly the flakes that occur on the surface must have come from eroded levels higher up in the soil profiles, and when flakes and sherds are mixed together on the surface the possibility that the two artefact categories co-existed at some time in the past cannot be ruled out. As we have seen, this is the case at many other sites. Secondly, the sampling and collecting strategies are not described. However it is clear that small shatter flakes were not included in his sample which tends to reduce the value of his metrical analysis. Finally, although he has a sample of more than 5,000 flakes to work with, for some reason he analyses only 2,982: he does not describe how this sample was selected nor does he indicate whether it included the excavated artefacts.

MINDANAO

Zamboanga and Sulu: Background

Spoehr (1973) has investigated two sites of interest in the southern Philippines—a site called Bungiao on the east coast of Zamboanga (ibid.: 73) and a cave site on Sanga Sanga Island in the Sulu Archipelago (ibid.: 106 ff.). Neither of these sites is described in detail.

There are two dates from layer 2 of the Sanga Sanga site, 6,650±180 B.P. and 7,945 ± 190 B.P. (ibid.: 55). The Sanga Sanga deposits consist of shell midden mixed with bone, stone stols and earthenware sherds. There are no metal or trade porcelains. Spoehr does not detail the faunal evidence from the site.

The Bungiao shelter contains burial jars, trade porcelains and other items, animal bones and some stone tools. It probably dates to about the 15th century A.D. There is some stratigraphy at this site and Spoehr suggests that since the trade procelains seem to be largely confined to the upper levels, the site may have been occupied before the introduction of the trade wares (ibid.: 166).

## Sanga Sanga Cave: Material Culture

### Stone Tools

Since there was only a test pit dug on the site, the resulting collections are small. The flakes are made from obsidian, tuff, silicified wood, andesite, and chert (Spoehr, 1973: 258 ff.). On the basis of the published evidence it appears that the industry has the following features:

* randomly shaped flakes;
* very little retouch;
* presence of at least one cobble tool;

and it superficially resembles the Tabon and LL/1 assemblages.

### Bone Tools

One bone tool, possibly an awl, was recovered.

### Shell Tools

The most interesting finds included a *Tridacna* gouge, a *Tridacna* shell adze, and three *Tridacna* tool blanks.

### Pottery

Some 462 sherds were recovered in the test excavation and Spoehr tentatively relates these to the Tabon Pottery Complex (ibid.: 185).

## Bungiao Rock Shelter: Material Culture

### Stone Tools

These are not described in detail. However, they are made from obsidian, chert and quartz, and appear to be unretouched amorphous flakes in the Tabonian tradition.

## Other Evidence

Shell bracelets, *Tridacna* shell pendants, glass beads and iron artefacts were recovered (ibid.: 242). The earthenware sherds are largely plain and undecorated, but some are cord marked, incised, check impressed and red slipped.

The shelter contained the remains of shells, pig, deer, monkey and snake. However, it is not clear whether these animals were brought into the cave by man.

## SITES IN OTHER AREAS

A blade tool tradition has been found in association with jar burial assemblages in Sorsogon and Albay Provinces on the islands of Luzon and Masbate, but they have not yet been described (Fox, 1967a: 107; Evangelista, 1967; Fox and Evangelista, 1957a; 1957b).

REFERENCES


Bellwood, P., 1978a. Archaeological research in Minahasa and the Talaud Islands, northeastern Indonesia. Asian Perspectives XIX: 240-288.

Bellwood, P., 1978b. Holocene flake and blade industries of Wallacea and their predecessors. Manuscript, Australian National University.

Bender, B., 1975. Farming in Prehistory: From Hunter-Gatherer to Food Producer. St. Martin's Press, New York.

Beyer, H. O., 1921. The Philippines before Magellan. Asia 21: 861-866; 924-928; 964; 968; 970.

Beyer, H. O., 1947. Outline review of Philippine archaeology by islands and provinces. Philippine Journal of Science 77: 205-374.

Cherry, R., 1978. An analysis of the lithic industry of Buad Island, Samar. Philippine Quarterly of Culture and Society 6: 3-80.

Chirino, P., 1604. Relation of the Filipinas Islands and of what has there been accomplished by the Fathers of the Society of Jesus. In: Blair, E. H. and J. A. Robertson (Eds.), The Philippine Islands 1493-1898. (trans.) Cacho Hermanos, Mandaluyong, Rizal, Volume 12: 175-321; Volume 13: 29-217.

Clarke, D. L., 1968. Analytical Archaeology. Methuen, London.

Colin, F., 1663. Native races and their customs In: Blair, E. H. and J. A. Robertson (Eds.), The Philippine Islands 1493-1898. (trans.) Cacho Hermanos, Mandaluyong, Rizal, Volume 40: 37-98.

Combes, J. F., 1667. The natives of the Southern Islands. In: Blair, E. H. and J. A. Robertson (Eds.), The Philippine Islands 1493-1898. (trans.) Cacho Hermanos, Mandaluyong, Rizal, Volume 40: 100-182.

Coutts, P. J. F. and J. P. Wesson, 1978a. Field reconnaissance of eastern Panay Island, Philippines. Philippine Quarterly of Culture and Society 6: 81-105.

Coutts, P. J. F. and J. P. Wesson, 1978b. Preliminary report on the Australian-Filipino archaeological expedition to Panay Island, January to March, 1978. Philippine Quarterly of Culture and Society 6: 299-306.

Coutts, P. J. F., Witter, D. and L. Cochrane, 1976. The archaeological evidence from the Cagayan Valley, northern Luzon: a review. Newsletter of the Far Eastern Prehistory Association 6: 12-29.

Dickerson, R. E., 1928. Distribution of Life in the Philippines. Monographs of the Bureau of Science, Manila.

Evangelista, A. E., 1962. Philippine archaeology up to 1950. Science Review 3: 17-22.

Evangelista, A. E., 1967. H. O. Beyer's Philippine Neolithic in the context of postwar discoveries in local archaeology. In: Zamora M. D. (Ed.), Studies in Philippine Anthropology. Alemar-Phoenix, Quezon City, pp. 63-87.

Fox, R. B., 1959. The Calatagan excavations: two 15th century burial sites in Batangas, Philippines. Philippine Studies 7: 321-390.

Fox, R. B., 1962. The Philippines in prehistoric times. Science Review 3: 1-16.

Fox, R. B., 1967a. Excavations in the Tabon Caves and some problems in Philippine chronology. In: Zamora, M. D. (Ed.), Studies in Philippine Anthropology. Alemar-Phoenix, Quezon City, pp. 88-116.

Fox, R. B., 1967b. The archaeological record of Chinese influences in the Philippines. Philippine Studies 15: 41-62.

Fox, R. B., 1968. Archaeology and the Philippines. Esso Silangan 18: 2-5.

Fox, R. B., 1970. The Tabon Caves. Monograph of the National Museum No. 1, Manila.

Fox, R. B., 1973. The Philippine Palaeolithic. Paper presented at the Conference on the Early Palaeolithic of East Asia, Montreal, August, 1973. Prehistory section of the IXth International Congress of Anthropological and Ethnological Sciences, Chicago, Illinois.

Fox, R. B., 1976. Notes on the stone tools of the Tasaday, gathering economies in the Philippines, and the archaeological record. In: Yen, D. E. and J. Nance (Eds.), Further Studies on the Tasaday. Panamin Foundation Research Series No. 2, pp. 3-12.

Fox, R. B. and Evangelista, A. E., 1957a. The Bato caves, Sorsogon Province, Philippines; a preliminary report of a jar burial-stone tool assemblage. Journal of East Asiatic Studies 6: 49-55.

Fox, R. B. and Evangelista, A. E., 1957b. The cave archaeology of Cagraray Island, Albay Province, Philippines: a preliminary report of explorations and excavations. Journal of East Asiastic Studies 6: 57-68.

Fox, R. B. and Peralta, J. T., 1974. Preliminaty report on the Palaeolithic archaeology of Cagayan Valley, Philippines, and the Cabalwanian industry Proceedings of the First Regional Seminar on South Asian Prehistory and Archaeology. (July-August 1972), Manila, pp. 100-147.

Fox, R. B., Peralta, J. T., Ewart, N., Paniza, I., Lopez, S., Cabanilla, I., Flores, R. and Aoyagi, Y., 1972. Ancient man and Pleistocene fauna in the Cagayan Valley, northern Luzon, Philippines: a progress report, Typescript, National Museum of the Philippines.

Gloria, M., 1939. A visit to the Negritos of Central Panay, Philippine Islands. Primitive Man 12: 94-102.

Glover, I. C., 1977. The Late Stone Age in eastern Indonesia. World Archaeology 9: 42-61.

Hole, F. and Heizer, R. F., 1973. An Introduction to Prehistoric Archaeology. Holt, Rinehart and Winston, New York, 3rd ed.

Hutterer, K., 1973. An Archaeological Picture of a Pre-Spanish Cebuano Community. San Carlos Publications, Cebu City.

Janse, O. R. T., 1944. Notes on Chinese influences in the Philippines in pre-Spanish times. Harvard Journal of Asiatic Studies 8: 34-62.

Jocano, F. L., 1967. Beyer's theory on Filipino prehistory and culture: an alternative approach to the problem. In: Zamora, M. D. (Ed.), Studies in Philippine Anthropology Alemar-Phoenix, Quezon City, pp. 128-150.

Jocano, F. L., 1975a. Philippine Prehistory: An Anthropological Overview of the Beginnings of Filipino Society and Culture. University of the Philippines System, Diliman, Quezon City.

Jocano, F. L. (Ed.), 1975b. The Philippines at the Spanish Contact. M.C.S. Enterprises, Manila.

Legaspi, A. M., 1974. Bolinao: A 14th-15th Century Burial Site. National Museum of the Philippines, Museum Publication No. 7, Manila.

McBryde, I., 1976. Seelands and Sai Yok pebble tools: a further consideration. Newsletter of the Far Eastern Prehistory Association 6: 46-61.

McCarthy, F. D., 1938. A comparison of the prehistory of Australia with that of Indo-China, the Malay Peninsula and the Netherlands East Indies. Proceedings of the Third Congress of Prehistorians of the Far East. Singapore.

McCarthy, F. D., 1941. Two pebble industry sites of Hoabinhian I type on the north coast of New South Wales. Records of the Australian Museum 21: 21-26.

Matthews, J. M., 1966. The Hoabinhian affinities of some Australian assemblages. Oceania 1: 5-22.

Morga, A. de, 1609. Events in the Filipinas Islands. In: Blair, E. H. and J. A. Robertson (Eds.), The Philippine Islands 1493-1898. (trans.). Cacho Hermanos, Mandaluyong, Rizal, Vol. 16, 27-29.

Movius, H. L., 1948. The Lower Palaeolithic cultures of Southern and Eastern Asia. The American Philosophical Society. Independence Square, Philadelphia 6, Vol. 38, Part 4.

Nance, J., 1975. The Gentle Tasaday: A Stone Age People in the Philippine Rain Forest. Harcourt, Brace Jovanovich, New York.

Peterson, W., 1974a. Summary report of two archaeological sites from north-eastern Luzon. Archaeology and Physical Anthropology in Oceania 9: 26-35.

Peterson, W., 1974b. Anomalous archaeological sites of northern Luzon and models of Southeast Asian prehistory. Unpublished Ph.D. thesis, University of Hawaii.

Plasencia, J. de., 1589. Customs of the Tagalogs. In: Blair, E. H. and J. A. Robertson (Eds.), The Philippine Islands 1493-1898. (trans.) Cacho Hermanos, Mandaluyong, Rizal, Vol. 7, 173-185.

Quirino, C. and Garcia, M., 1958. The manners, customs, and beliefs of the Philippine inhabitants of long ago: being chapters of 'A late 16th century Manila manuscript', transcribed, translated and annotated. Philippine Journal of Science 84: 325-449.

Rahmann, R., 1955. Bow and arrow of the Visayan Negritos. Philippine Journal of Science 84: 323-333.

Rahmann, R., 1958. Some notes on the Negritos of Iloilo, Island of Panay, Philippines. Studia Instituti Anthropos 53: 864-876.

Rahmann, R., 1963. The Negritos of the Philippines and the early Spanish missionaries. Studia Instituti Anthropos 18: 137-157.

Rahmann, R. and Maceda, M. N., 1955. Notes on the Negritos of northern Negros. Studia Instituti Anthropos 50: 810-837.

Rahmann, R. and Maceda, M. N., 1962. Notes on the Negritos of Antique, island of Panay, Philippines. Studia Instititi Anthropos 57: 626-643.

Ronquillo, W. P., 1977. Report on the Preliminary Archaeological Excavations of Rabel Cave. Typescript, National Museum of the Philippines, Manila.

San Antonio, J. F. de, 1738. The native peoples and their customs. In: Blair, E. H. and J. A. Robertson (Eds.), The Philippine Islands 1493-1898. (trans.) Cacho Hermanos, Mandaluyong, Rizal, Vol. 40, 296-373.

Santaren, Fr., 1956. Bisayan accounts of early Bornean settlements in the Philippines. Sarawak Museum Journal 7: 22-42.

Sieveking, A., 1958. The Palaeolithic industry of the Kota Tampan, Perak, northwestern Malay. Asian Perspectives 2: 91-102.

Solheim, W. G., 1952a. Oceanian pottery manufacture. Journal of East Asiatic Studies 1: 1-39.

Solheim, W. G., 1952b. Paddle decoration of pottery. Journal of East Asiatic Studies 11: 35-45.

Solheim, W. G., 1952c. Pottery manufacturing in the islands of Masbate and Batan, Philippines. Journal of East Asiatic Studies 1: 49-53.

Solheim, W. G., 1954. Ibanag pottery manufacture in Isabela, Philippines. Journal of East Asiatic Studies 3: 305-307.

Solheim, W. G., 1955. Notes on the archaeology of Masbate. Journal of East Asiatic Studies 4: 47-50.

Solheim, W. G., 1958. The Kalanay Pottery Complex in the Philippines. Artibus Asiae 20: 279-288.

Solheim, W. G., 1959a. Introduction to Sa-Huynh. Asian Perspectives 3: 97-108.

Solheim, W. G., 1959b. Further notes on the Kalanay Pottery Complex in the Philippines. Asian Perspectives 3: 157-165.

Solheim, W. G., 1964a. The Archaeology of the Central Philippines: A Study Chiefly of the Iron Age and its Relationships. Monograph of the National Institute of Science and Technology No. 10, Manila.

Solheim, W. G., 1964b. Pottery and the Malayo-Polynesians. Current Anthropology 5, 360: 376-384.

Solheim, W. G., 1967. The Sa-Huynh-Kalanay Pottery Traditions: past and future research. In: Zamora, M. D. (Ed.), Studies in Philippine Anthropology. Alemar-Phoenix, Quezon City, pp. 151-174.

Solheim, W. G. and Shuler, T., 1959. Further notes on Philippine pottery manufacture: Mountain Province and Panay. Journal of East Asiatic Studies 8: 1-10.

Spoehr, A., 1973. Zamboanga and Sulu: an archaeological approach to ethnic diversity. Ethnology Monographs No. 1.

Tenazas, R. C. P., 1973. Excavations in Victorias, Negros occidental: a brief report. Philippine Quarterly of Culture and Society 1: 274-277.

Thiel, B., n.d. Excavations in the Pinacanavan Valley, Northern Luzon, Unpublished Manuscript.

Tindale, N. B., 1937. Relationship of the extinct Kangaroo Island culture with cultures of Australia, Tasmania and Malaya. Records of the South Australian Museum 6: 39-60.

Tuggle, H. D. and Hutterer, K. L., 1972. Archaeology of the Sohoton area, south-western Samar. Leyte-Samar Studies 1(2): 1-104.

Wernstedt, F. L. and Spencer, J. E., 1967. The Philippine Island World. University of California Press, Berkeley.

Willey, G. R. and Phillips, P., 1958. Method and Theory in American Archaeology. University of Chicago Press, Chicago.

Yengoyan, A. A., 1967. The initial populating of the Philippines: some problems and interpretations. In: Zamora, M. D. (ed.), Studies in Philippine Anthropology, Alemar-Phoenix, Quezon City, pp. 175-185.

# 7. THE PREHISTORIC EXPLOITATION OF STONE RESOURCES IN NORTHERN NEW ZEALAND

Janet Davidson

Two lithic resources have long been recognized as extremely important to the prehistoric inhabitants of New Zealand. These were nephrite (which occurs only in the south and west of the South Island (see Figure 7.1) and obsidian (which is found at a number of locations in the northern half of the North Island (see Figure 7.3). Both are the subject of other contributions to this symposium (Beck; Leach and Warren; this volume). The significance and wide distribution of these two resources were appreciated by nineteenth century scholars, and various versions of Maori oral traditions personifying the two types of rock were recorded (Duff, 1956: 237-239).

It is now known that although both these materials were distributed throughout New Zealand in prehistoric times, many other rock types were also transported over long distances. Foremost among these were certain fine-grained rocks used for stone adzes, of which the most notable are the metasomatised argillites of the Nelson mineral belt in the northern South Island (Keyes, 1975; Walls, 1974). A series of prehistoric quarry and working sites have been identified in this region, some of the products of which reached the extremes of both islands (See Figure 7.1). Other adze-making centres are also known. However, the full extent of Maori exploitation of stone resources is only now beginning to be demonstrated by the accurate sourcing of the full range of stone material recovered in excavations. The first results of this approach, applied to a series of excavations at Palliser Bay in the southern tip of the North Island, revealed a complex network of communications extending a very considerable distance both north and south of the area under consideration and involving many different kinds of rock (Prickett, 1975; 1979; Leach, 1978; Leach and Anderson, 1978).

This paper is concerned with the northern half of the North Island—the area within which the known sources of obsidian occur. Although sourcing studies are still in their infancy, sufficient data already exist to indicate the complexity of prehistoric communication as reflected by the distribution of stone resources.

## ADZES

Two early adze manufacturing centres are known in the northern North Island, one of undoubtedly major significance, the other apparently of more minor and local significance. The Tahanga basalt source is located on the east coast of the Coromandel Peninsula, and Motutapu greywacke on a small island near Auckland City (Figure 7.2).

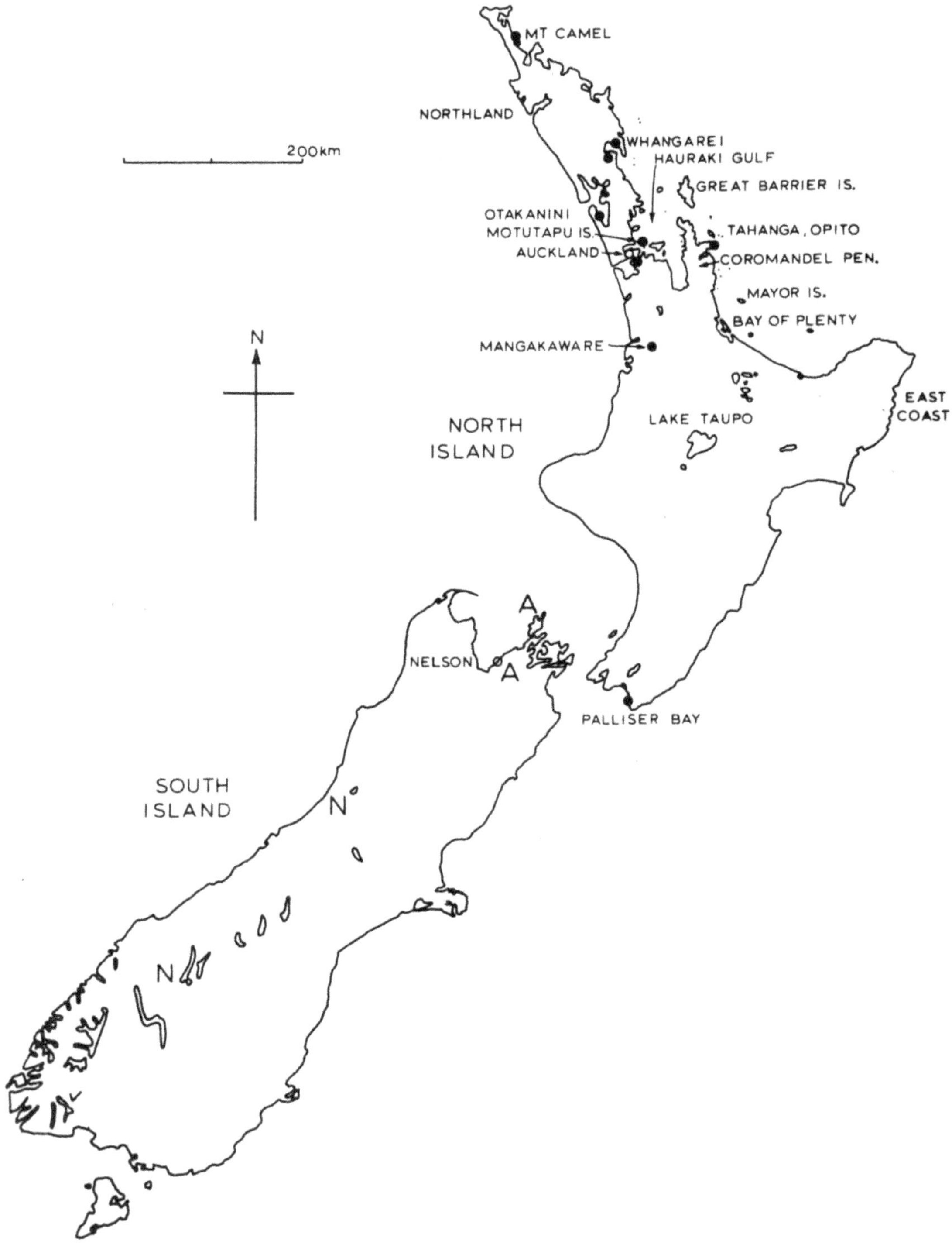


Figure 7.1 Map of New Zealand showing places mentioned in the text. N = nephrite, A = metasomatised argillite.

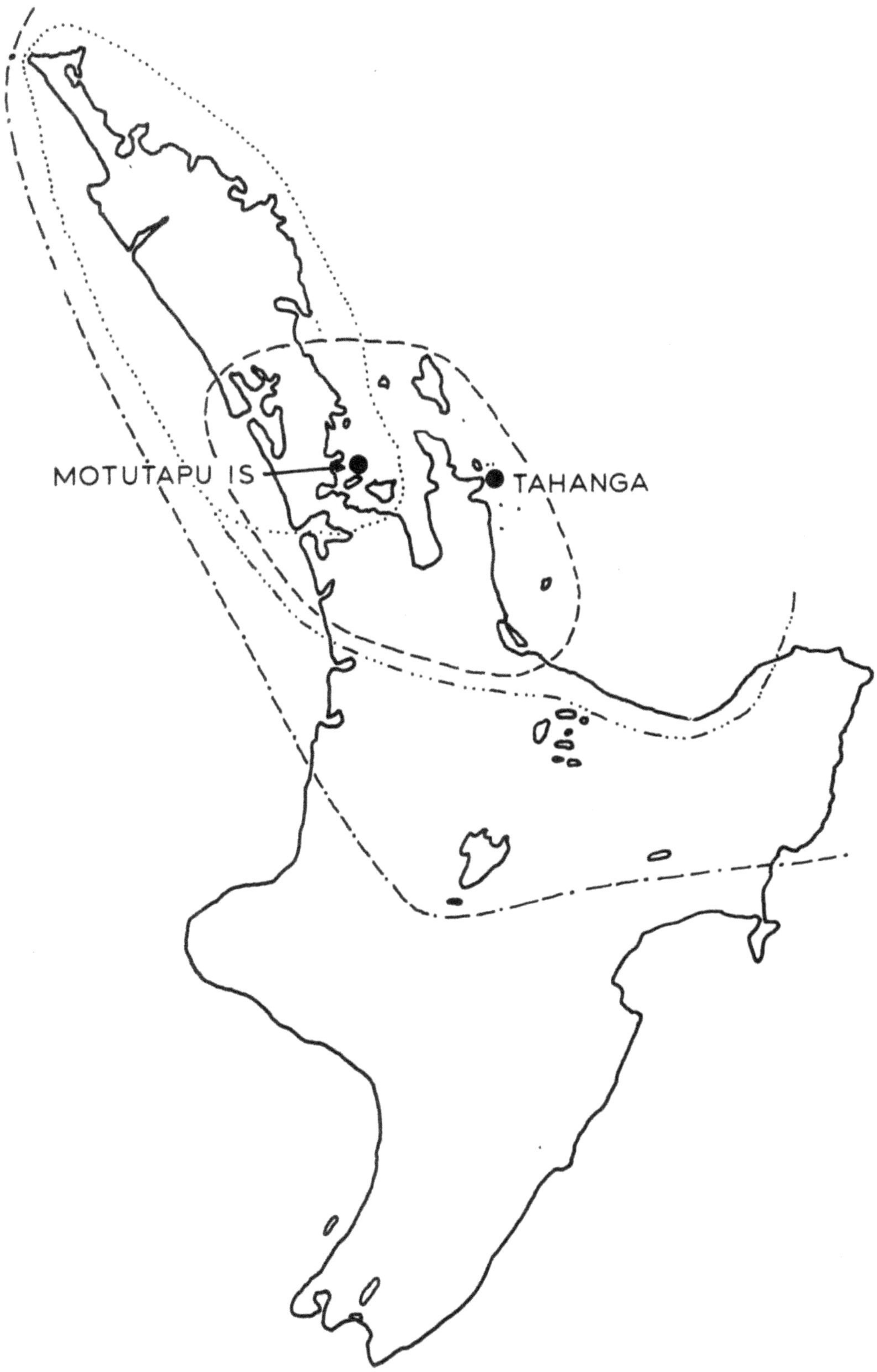


Figure 7.2 The positions of the Tahanga and Motutapu adze manufactories, and the distribution of adzes in Tahanga basalt and Northland gabbro (after Moore, 1975 and Best, 1977).

| | |
|---|---|
| — — — — — | Tahanga adzes numerous |
| — — · — — · | Occasional Tahanga adzes |
| ............ | Gabbro adzes numerous |
| — ··· — ··· | Occasional Gabbro adzes |

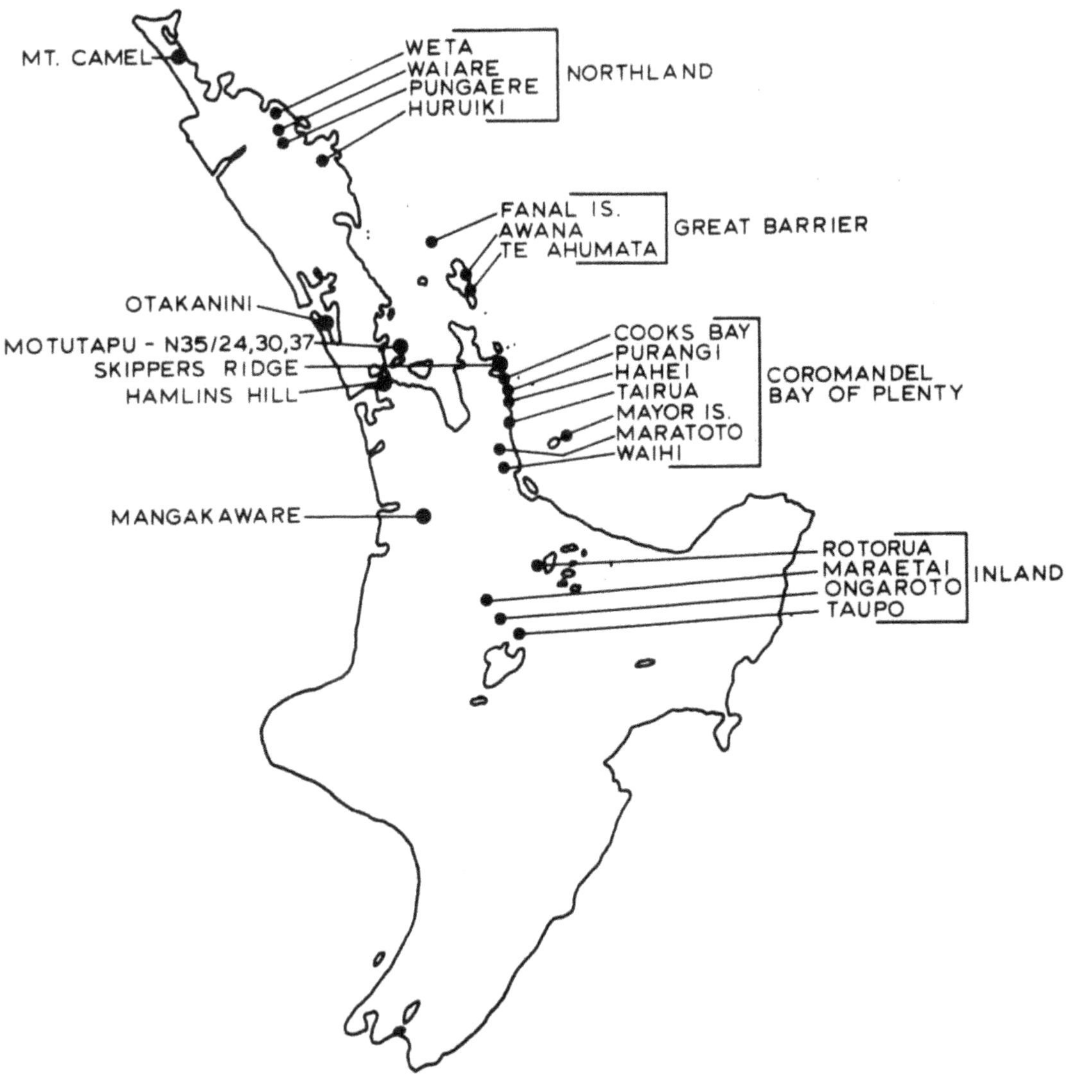


Figure 7.3 New Zealand Obsidian sources (after Leach and Anderson, 1978), and northern archaeological sites from which obsidian has been sourced.

Tahanga basalt has been studied in some detail (Moore, 1975; 1976; Best, 1975; 1977). The stone is particularly fine-grained and is apparently unique to the small area that was exploited on Tahanga Hill. Stone was worked at several places on the hill itself and preforms were transported to a number of places within a radius of approximately 60 km of the source for further processing and finishing. Completed products are numerous within 150 km of the source and scattered examples are found within a radius of 350 km and possibly much further afield. Best has shown that certain early sites far from the source have a large number of adzes of Tahanga basalt, most notably the Mt Camel site more than 300 km to the north-west.

The Tahanga basalt source was discovered early in New Zealand prehistory, perhaps about A.D. 1100, and probably continued to flourish until about A.D. 1500. Thereafter it may have been of much more local significance, although it was quite possibly still in use at the time of Cook's visit to nearby Mercury Bay in A.D. 1769.

Motutapu greywacke is much less studied than Tahanga basalt. Exposures of suitable rock outcrop at a number of places around the coast of the island. Suitable pieces were taken to adjacent beach sites and worked into preforms and finished adzes. The material was exploited from the earliest known settlement of the island in the fourteenth century until the eighteenth century. During this time the forms of adze made changed from a range of Archaic types typical of early sites to a more restricted form representing a regional style typical of late northern prehistory (Davidson, 1972; 1978).

Motutapu greywacke belongs to the Waipapa Group of probably Jurassic Age (Mayer, 1968; 1969), which outcrops in many places in Northland as well as on the Hauraki Gulf islands such as Motutapu, and the adjacent south Auckland mainland. It is by no means certain that prehistoric exploitation of this geological formation was confined to Motutapu Island. On the contrary, it is probable that other working areas existed, particularly on the adjacent islands. The extent of distribution of tools in this material has not been studied, although adzes which may well originate from this source have been found over an area comparable to that within which Tahanga basalt adzes are found.

The use of Motutapu greywacke is best understood as it relates to prehistoric life on Motutapu itself. It appears that certain localities were used as manufactories for the initial preparation of preforms over many centuries. At many of the small hamlets which also existed on the island, individual craftsmen worked on the fine finishing of tools, probably for their own and their immediate relatives' use (Leahy, 1970; 1972).

A significant feature of adze manufacture and distribution, which has not been widely appreciated, is the extent to which workers engaged in adze manufacture in the vicinity of major quarries also owned, used, and modified products from rival centres. At one manufacturing site on Motutapu Island (Pig Bay), a complete small adze in metasomatised argillite, almost certainly from the South Island, was found among a profusion of preforms of local greywacke; at another (Sunde Site), a preform of Tahanga basalt was associated with preforms of local rock (Moore, 1976: 81). Yet more significantly, at the Opito Beach Site almost under the shadow of Tahanga, where numerous preforms of Tahanga basalt were found, three small complete adzes of greywacke were also recovered. Moreover, there were flakes derived from finished tools of more than one variety of South Island metasomatised argillite in the same deposit, even though no finished tools in this material were found (Boileau, in press). Similar flakes derived from metasomatised argillite adzes have been found in other Coromandel sites where Tahanga basalt predominates.

These examples show, firstly, that artisans at Tahanga and elsewhere were in at least indirect contact with each other, through possession of each other's products, and secondly, that adzes from the South Island, which from surface finds are known to have penetrated throughout the North Island, were actually also reaching the northern quarrying and manufacturing sites. Ideas about desirable forms, if not information about actual manufacturing techniques, could have travelled throughout the country in this way.

Stone adzes were undoubtedly made in many other rocks and in many other centres in the northern North Island than the two so far mentioned, particularly during later times. Best has demonstrated (1975; 1977) that adzes made of a Northland gabbro have a distribution in the northern North Island not far inferior to that achieved by Tahanga basalt (Figure 7.2). The exact source of the gabbro is not known, and the period of its use is uncertain, but it is presumed to be late and to belong to the period when Tahanga basalt was largely restricted to local usage. Best's study of the gabbro was confined to surface finds in museum collections, and documentation of this rock type and its associations by excavation is urgently needed.

Other rocks used for adzes included andesites and sandstones. Greywacke was widely used in later times, but there are many other possible sources of greywacke in the northern North Island besides the Waipapa Group already mentioned. The sources and distribution of adzes in all these materials present an urgent problem for further investigation. It should also be noted that just as South Island metasomatised argillite adzes were reaching the furthest north, particularly in early times, so the most valued of all adzes, those of South Island nephrite, also reached the northern North Island in significant numbers in later times.

OBSIDIAN

Obsidian is probably the most widely distributed and frequently found rock in New Zealand archaeological sites, considering the country as a whole. According to ethnographic accounts, it was used for kitchen knives, surgical instruments, and for cutting hair, bodies, flax and raupo. Conveniently sized blocks could be seen in all Maori settlements in the early nineteenth century and were kept for general use, and as important articles of barter (Maning, 1876).

Characterization of sources has been investigated in detail, although sourcing of large quantities of archaeological specimens is still an ideal rather than a reality. Ward (1972; 1974) listed 28 major discrete localities from which he had obtained obsidian samples and which he regarded as at least potential sources of artefacts. These were reduced to 18 petrographically distinct sources in 8 major areas: Northland, Great Barrier Island, Fanal

Island, Coromandel Peninsula, Rotorua, Maraetai, Taupo, and Mayor Island. More recently, Leach and Anderson (1978) in their analysis of obsidian from archaeological sites in Palliser Bay (far to the south of the source areas) grouped the sources into four geographic regions, namely Northland, Great Barrier, Coromandel-Bay of Plenty (including Mayor Island), and Inland (Figure 7.3). The question of characterization of obsidian sources is further discussed in this symposium (Leach and Warren, this volume).

In the following discussion, the prehistoric exploitation of obsidian in New Zealand is considered from two points of view—the dissemination of obsidian beyond the region within which all known sources occur, and distribution within that region. The most detailed studies to date have been carried out beyond the source region, but all indications are that distribution within the region is also extremely complex.

The twelfth century inhabitants of Palliser Bay were obtaining obsidian from Coromandel, Mayor Island and Inland sources. By the fourteenth century, Northland obsidian was also reaching Palliser Bay. The presence of Northland obsidian in Palliser Bay is particularly surprising, since it must have overtaken material from most other sources during its long journey down the island. It is possible, of course, that Northland obsidian was passed down the west coast, where its only competitor might be Inland material being sent down one of the two major rivers that flow from the central North Island to the west coast. At the southern end of the North Island it would meet and mingle with obsidian from other sources coming down the east coast. The recipients at Palliser Bay would probably have little or no idea where any of their obsidian had originally come from. These are, of course, only speculations until much more obsidian is sourced and the variations, if any, between east coast and west coast distributions defined.

In the Far North of the North Island, the early site of Mt Camel is also beyond the obsidian source regions, although it is not far from the most northerly sources. Over 90 percent of the obsidian in this site is from Mayor Island, although there is definitely a small amount from the nearest Northland sources, and some from at least one other source again (Best, 1977: 318; Best, pers. comm.). The surprisingly high proportion of Mayor Island obsidian correlates with the presence in this particular site of other stone from the Coromandel-Bay of Plenty region—Tahanga basalt and a siliceous sinter mentioned below.

Beyond the obsidian-producing region, then, people were receiving a mixture of obsidian from a variety of sources. Such slight data as are available suggest that this was also the case within the obsidian source region. Of seven sites for which some obsidian source identifications have been published, five are said to have material from at least three sources, while the remainder have two (Reeves and Ward, 1976: 280; see also Davidson, 1970 for Hamlins Hill in Auckland). On Motutapu Island, for example, obsidian was apparently coming from Mayor Island, Great Barrier Island, Northland (Huruiki), and Coromandel. The extremely small samples, and the tentative and preliminary nature of much of this work, mean that these results can only be taken as a very preliminary indication of the sort of patterns that may be able to be identified in future. It is almost certain that more detailed and precise work on large numbers of specimens would show that most communities were obtaining obsidian from more sources rather than from fewer.

The distribution of obsidian in some ways resembles that of adzes. Although there was a concentration of products of a particular source within a defined radius of that source, some pieces travelled very long distances beyond their source area. Just as products from different adze manufactories and from the South Island were distributed throughout the northern regions, so some obsidian was passed right through the vicinity of competing sources to end up far from its point of origin.

The old belief that all New Zealand obsidian came from one source—Mayor Island—can be partly understood, if not justified, by the apparent pre-eminence of Mayor Island obsidian throughout New Zealand prehistory. No archaeological site in New Zealand has yet been shown not to have obsidian from Mayor Island, even though the proportion of Mayor Island obsidian to that from other sources cannot be used as an indication of age, as was once hoped (Leach and de Souza, 1979). Most sites are thought to have obsidian from Mayor Island and one or more other sources. Its continuing importance is reflected both in the early Mt Camel site, and in the fact that northern warriors were believed to have brought back blocks of obsidian from Mayor Island, rather than other sources, during their southern raids in the early nineteenth century (Maning, 1876).

OTHER ROCKS

In a recent review of New Zealand cherts, Moore (1977) adopted a broad definition of chert to include various siliceous rocks commonly found in archaeological sites. He identified the Northland-Auckland-Coromandel region as one of three main zones in the country within which good flake quality chert occurs. Twenty-five known sources of flake quality chert were listed in this zone (Figure 7.4) and the extent of potential chert-bearing formations suggests the likelihood that many more exist. Many of these chert sources are close to the other important stone resources previously mentioned—the obsidian sources of Northland, Great Barrier Island and the Coromandel Peninsula, and the Tahanga and Motutapu adze manufactories. People living in these regions were particularly well supplied with stone compared with those of some other regions. Further chert sources exist in the eastern Bay of Plenty-East Coast area, and there are others in the south-eastern North Island and in various parts of the South Island.

Best has suggested that the chert-like material used for drill points at Mt Camel (which he identifies as siliceous sinter) almost certainly came from the same part of the Cormandel Peninsula as Tahanga, more than 300 km from the site (Best, 1977: 318). This is the first time that material of this kind has been shown to have been widely transported beyond its point of origin. Previously, use of the Coromandel cherts appeared to be confined to the Coromandel Peninsula, where they are often found in the same sites at which adzes were actually being produced from Tahanga basalt. It remains to be seen whether Mt Camel is unique in this respect, or whether Coromandel chert will be found in other remote sites.

Figure 7.4 Northern New Zealand chert sources (after Moore, 1977).

Other sources of chert appear (on present evidence) to have been exploited on a very restricted scale compared with obsidian. Chert flakes have been found in some quantity at two sites near Whangarei: Ruarangi (Hougaard, 1971) and the Northland Harbour Board site (A. Walton, pers. comm.), both relatively recent in age. This chert probably comes from one or more of the several known sources in the immediate vicinity of Whangarei. Other chert sources, for example, elsewhere in Northland, and in the Eastern Bay of Plenty-East Coast region, probably had equally restricted spheres of influence.

Prehistoric New Zealanders used stone for many other purposes besides making adzes, knives, and drill points. Items such as hammer stones, pounders, fishing sinkers, grinders and files may also in some cases have been obtained from a distance, but are largely unstudied. Hammers used in adze manufacture at Motutapu appear to be of local origin; pebbles of the cherts that are interbedded with greywacke and argillite in the Waipapa Group were often used. At Tahanga, however, the large hammers that were used in the initial working of basalt blocks were themselves imported from the other side of the Coromandel Peninsula, some 40 km in a direct line, and considerably more by the coastal route presumably used.

The stone fighting clubs (patu onewa) that were such a feature of Maori society at the end of the prehistoric period present an interesting problem for future study. Although they have not been investigated in any detail it appears that these weapons were made from some of the same kinds of stone, such as greywacke and sandstone, that were also used for adzes in later times, and may have originated from the same sources. The development of these weapons may well be related to the decline of resources such as Tahanga basalt, and the rise of adzes in materials such as the gabbro studied by Best.

## THE RANGE OF RESOURCES IN INDIVIDUAL SITES

The study of the distribution of products from a single source, while very instructive, must be matched by the identification of all stone resources from particular occupation sites, as was done in Palliser Bay (Prickett, 1975; 1979), if the full extent of prehistoric communication is to be appreciated. Studies of this kind are still in their infancy.

In a pioneering study of artefacts found at Otakanini, near Auckland, Bellwood (1972: 283-6) showed that the inhabitants of this relatively late site were using greywacke for fighting clubs, andesite for adzes, sandstone for grinders, baked mudstone and chert for flake tools, and dacite for files. All of these could have been obtained within 50 km of the site, and most from considerably closer. Moreover most sources were at least partly accessible by canoe. Obsidian provided the exception, apparently coming from four different sources all 100 km or more from the site. This pattern is rather similar to that found in late sites on Motutapu Island where local stone was used for adzes, hammers, grinders and some flake tools, but obsidian and nephrite indicate a network of contacts extending north, east, and south-east and ultimately to the South Island (Davidson, 1978: 333).

At an inland site at Mangakaware dated to the late seventeenth or eighteenth centuries, Bellwood (1978: 38-40) found that the majority of stone artefacts, including adzes, fighting clubs, and pounders, were made from Mesozoic sandstones and argillites which could have been obtained within 12 km of the site. Obsidian was obtained from at least two sources, and more surprising imports were a large flake of siliceous sinter, from either Rotorua-Taupo or the Coromandel region (which could have travelled the same routes as the obsidian), and three pieces of quartzite, for which a possible South Island source was suggested.

## CONCLUSION

In so far as patterns can yet be detected in the northern North Island, it appears that at both early and late sites, a majority of stone resources often came from within about 50 km of the site, and sometimes much less. Obsidian is a regular exception, coming from up to 300 km away and present in almost all sites. Surprisingly frequent under the circumstances are South Island imports, particularly metasomatised argillite and nephrite, but possibly also other stones as well. Mt Camel, with its amazing dependence on remote resources, is at present a unique exception to this general pattern, but future investigation of other early sites in the region may yet show that it is not the isolated case it now appears to be.

## REFERENCES


Bellwood, P., 1972. Excavations at Otakanini Pa, South Kaipara Harbour. Journal of the Royal Society of New Zealand 2(3): 259-291.

Bellwood, P., 1978. Archaeological research at Lake Mangakaware, Waikato, 1968-1970. University of Otago studies in prehistoric anthropology Vol. 12, and New Zealand Archaeological Association Monograph 9.

Best, S. B., 1975. Adzes, rocks and men. Unpublished M.A. research essay, Anthropology Department, University of Auckland.

Best, S. B., 1977. The Maori adze: an explanation for change. Journal of the Polynesian Society 86(3): 307-337.

Boileau, J., in press. The artefact assemblage from the Opito Beach Midden, N40/3, Coromandel Peninsula. Records of the Auckland Institute and Museum 17.

Davidson, J. M., 1970. Salvage excavations at Hamlins Hill, N42/137, Auckland, New Zealand. Records of the Auckland Institute and Museum 7: 105-122.

Davidson, J. M., 1972. Archaeological investigations on Motutapu Island, New Zealand. Introduction to recent fieldwork, and further results. Records of the Auckland Institute and Museum 9: 1-14.

Davidson, J. M., 1978. The prehistory of Motutapu Island, New Zealand. Five centuries of Polynesian occupation in a changing landscape. Journal of the Polynesian Society 87(4): 327-337.

Duff, R., 1956. The moa-hunter period of Maori culture, 2nd edition. Government Printer, Wellington (Canterbury Museum Bulletin 1).

Hougaard, M. P., 1971. Excavations on Ruarangi Pa (Site N20/41), Whangarei, New Zealand. Records of the Auckland Institute and Museum 8: 1-22.

Keyes, I. W., 1975. The D'Urville Island - Nelson metasomatised rocks and their significance in New Zealand prehistory. Historical Review 23(1): 1-17.

Leach, B. F., 1978. Four centuries of community interaction and trade in Cook Strait, New Zealand. Mankind 11(3): 391-405.

Leach, B. F. and Anderson. A. J., 1978. The prehistoric sources of Palliser Bay obsidian. Journal of Archaeological Science 5: 301-307.

Leach, B. F. and de Souza, P., 1979. The changing proportions of Mayor Island obsidian in New Zealand prehistory. New Zealand Journal of Archaeology 1: 29-51.

Leahy, A., 1970. Excavations at Site N38/30, Motutapu Island, New Zealand. Records of the Auckland Institute and Museum 7: 61-82.

Leahy, A., 1972. Further excavations at Site N38/30, Motutapu Island, New Zealand. Records of the Auckland Institute and Museum 9: 15-26.

Maning, F. E., 1876. Extracts from a letter from F. E. Maning, Esq., relative to the extinction of the moa. Transactions of the New Zealand Institute 8: 102-103.

Mayer, W., 1968. The stratigraphy and structure of the Waipapa Group on the islands of Motutapu, Rakino, and the Noisies Group near Auckland, New Zealand. Transactions of the Royal Society of New Zealand, Geology 5(9): 215-233.

Mayer, W., 1969. Petrology of the Waipapa Group, near Auckland, New Zealand. New Zealand Journal of Geology and Geophysics 12: 412-435.

Moore, P. R., 1975. Preliminary investigation of the Tahanga basalt, Coromandel Peninsula. New Zealand Archaeological Association Newsletter 18(1): 32-36.

Moore, P. R., 1976. The Tahanga basalt: an important stone resource in North Island prehistory. Records of the Auckland Institute and Museum 13: 77-93.

Moore, P. R., 1977. The definition, distribution, and sourcing of chert in New Zealand. New Zealand Archaeological Association Newsletter 20(2): 51-85.

Prickett, K. E., 1975. The prehistoric exploitation and knowledge of geological resources in Southern Wairarapa. Unpublished M.A. thesis, Anthropology, University of Otago.

Prickett, K. E., 1979. The stone resources of early communities in Palliser Bay. In: Leach, B. F. and H. M. Leach (Eds.), Prehistoric Man in Palliser Bay, pp. 163-184. National Museum of New Zealand Bulletin 21.

Reeves, R. D. and Ward, G. K., 1976. Characterization studies of New Zealand obsidians: toward a regional prehistory. In: Taylor R. E. (Ed.), Advances in obsidian glass studies, archaeological and geochemical perspectives, pp. 259-287. Noyes Press, New Jersey.

Walls, J. Y., 1974. Argillite quarries of the Nelson mineral belt. New Zealand Archaeological Association Newsletter 17(1): 37-43.

Ward, G. K., 1972. Obsidian and New Zealand archaeology: a paradigm for sourcing artefact assemblages using X-ray fluorescence spectrography. Unpublished M.A. thesis, Anthropology, University of Otago.

Ward, G. K., 1974. A paradigm for sourcing New Zealand archaeological obsidians. Journal of the Royal Society of New Zealand 4(1): 47-62.

# 8. PROVENANCES OF SANUKITE IMPLEMENTS IN WESTERN JAPAN

T. Higashimura, T. Warashina and Y. Kamaki

A geochemical rule that a rock has a unique pattern of element composition which is characteristic of its petrographic province can be applied to determine the source of stone artefacts. Using this principle, neutron activation analysis has been applied to obsidian implements in North America (Gordus et al., 1967) and many other districts (De Bruin et al., 1972; Aspinall and Feather, 1972). Fission track analysis of obsidian has been used by Suzuki (1973) to estimate its uranium content and the date of solidification of the obsidian flow. These two variables allow one to establish the provenance of obsidian implements.

X-ray fluorescence analysis is also a useful method for this purpose. The technique is so simple that archaeologists who are untrained in analysis can carry it out. Stone samples can be analysed without destruction, if we use the element-ratio method which will be discussed later. Many elements can be analysed at the same time with the energy-dispersive X-ray fluorescence spectrometer, and this economical use of time is very important, because many samples must be analysed to obtain a clear conclusion on trade routes in ancient times.

Several works have been published on problems of allocating stone artefacts to sources after employing X-ray fluorescence analysis; however, these have been limited to obsidian (Ward, 1974b; Nelson et al., 1975; Smith et al., 1977). In this paper, X-ray fluorescence analysis is applied to ancient sanukite implements.

Figure 8.1 is a map of Japan which illustrates regions of different stone utilization in the stone age. In northern Japan, hard shale was used for implements. In the northeast and southeast parts of Japan, obsidian was the material which the most frequently used for stone implements. In the western part of Japan, more than 90% of the stone implements were made from sanukite. This sanukite area is an important one in the history of Japan. After the fifth century A.D., this was the most civilized area in Japan. For several hundred years before this, archaeological evidence points to the existence of two earlier civilized areas which are included in the sanukite area. However, in the Jomon period, which continued for several thousand years before the third century B.C., there might have been many regional cultural centres in western Japan. Food gatherers in this period would have been wide ranging, and there would have been strong interaction between people in distant localities. An important question is how far these people ranged to trade or gather their resources in this period. Provenance studies on sanukite implements may provide information to help answer this question.

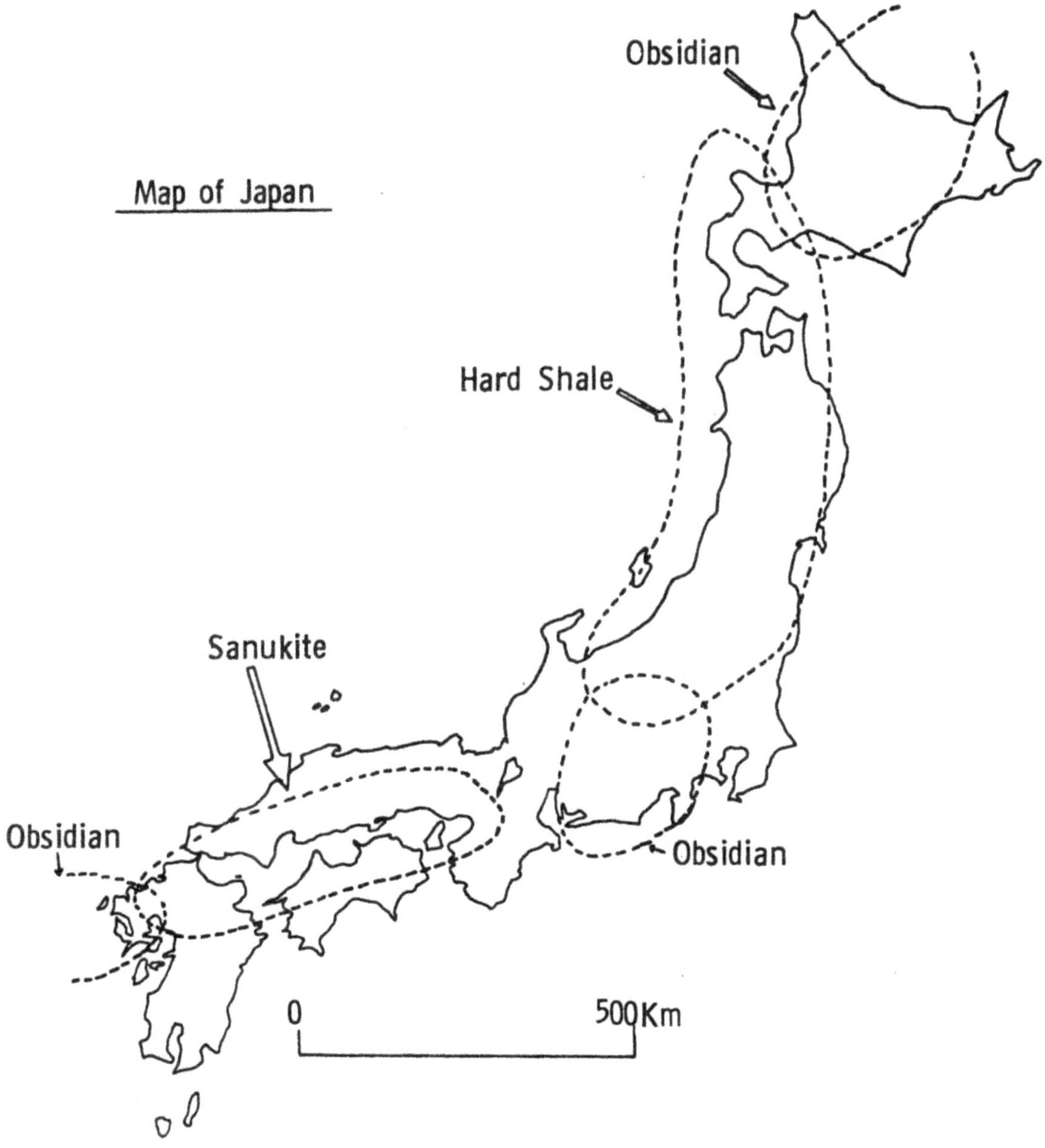


Figure 8.1 Regions of different stone utilization in the stone age.

Sanukite is one of the hyperthene andesites. It was extruded about 15 million years ago. It is black in colour and very hard. It is flaked by conchoidal fracture, producing very sharp cutting edges. In western Japan, there is no geological source of obsidian, but there are several rich sources of sanukite. Sanukite was used frequently in the Palaeolithic and Neolithic periods in Western Japan. Although there are more than a dozen different source localities of sanukite, only two of them have been noted by archaeologists. Until recently it was thought that all sanukite implements were transported, probably by trade, from one or other of these two sources.

The X-ray fluorescence spectrometer used for this study consists of a radioactive isotope source, a silicon detector, a multichannel pulse height analyser, a paper tape punch, and an automatic sample changer. Three elements in a sample, Si, K and Ca, can be analysed with a 50 mCi $^{55}$Fe radioactive isotope source; and seven elements, Ti, Mn, Fe, Rb, Sr, Y, Zr and Nb, can be analysed with a 10 mCi $^{109}$Cd radioactive isotope source. The sample holder is made from a circular plastic cylinder 40 mm in diameter and 3 mm in height with 3 mil Milar film at the bottom. The sample is located on this film and is excited from the bottom with the X-rays of the radioactive source. Twenty-four samples are contained in the sample changer and are analysed automatically within a full day. X-ray fluorescence spectra of these 24 samples are recorded in a paper tape, and areas of fluorescence X-ray peaks of each element are later obtained with a computer fitting procedure (Higashimura and Warashina, 1975; Warashina et al., 1978).

For the analysis of archaeological objects, it is very important to employ a non-destructive method. X-ray fluorescence analysis is ideally suited because the fluorescence intensity can be measured without destroying the object. A major difficulty in non-destructive analysis is eliminating the effects of differences in shape and size of samples. These differences influence the fluorescence peak area of each element by approximately the a same fraction. So, if we take ratios of peak areas of each element to the peak area of a reference element, the shape effect can be largely eliminated. Reference elements were Ca in the case of the $^{55}$Fe radioactive source and Sr in the case of the $^{109}$Cd radioactive source.

Table 8.1 shows an example of values of peak-ratios of sanukite samples taken from a geographical source. The first row shows the average values of powder samples made from 14 different blocks. The bottom row shows those of chip samples of different shapes. The consistency between the powder samples and the chip samples is very good. Differences between the two types of sample are less than the standard deviations of results for any one source. Thus, the shape effect was largely eliminated by taking these peak-ratios.

The first step in this study was to obtain data on peak-ratios of sanukite stones taken from different geographical sources; in other words, to establish finger prints of sanukite stones from the different sources. There are about twenty geographical sources of this stone in Western Japan as is shown in Figure 8.2. Both open circles and filled circles in this figure show geographical sources of sanukite. We did not find any archaeological implements which were made from sanukite stones taken from sources designed by open circles, but we found many implements which were made from stones taken from sources designated by filled circles.

Each 14 sanukite samples taken from a locality

| | K/Ca | Ti/Ca | Rb/Sr | Y/Sr | Zr/Sr |
|---|---|---|---|---|---|
| Powder | 0.282 ±0.008 | 0.234 ±0.005 | 0.232 ±0.015 | 0.051 ±0.011 | 0.547 ±0.042 |
| Chips of different shape | 0.284 ±0.008 | 0.239 ±0.007 | 0.226 ±0.016 | 0.059 ±0.024 | 0.563 ±0.035 |

Table 8.1 Peak-ratios of chip samples and powder samples of sanukite stones taken from the Nijohsan source. In each case 14 samples were analysed.

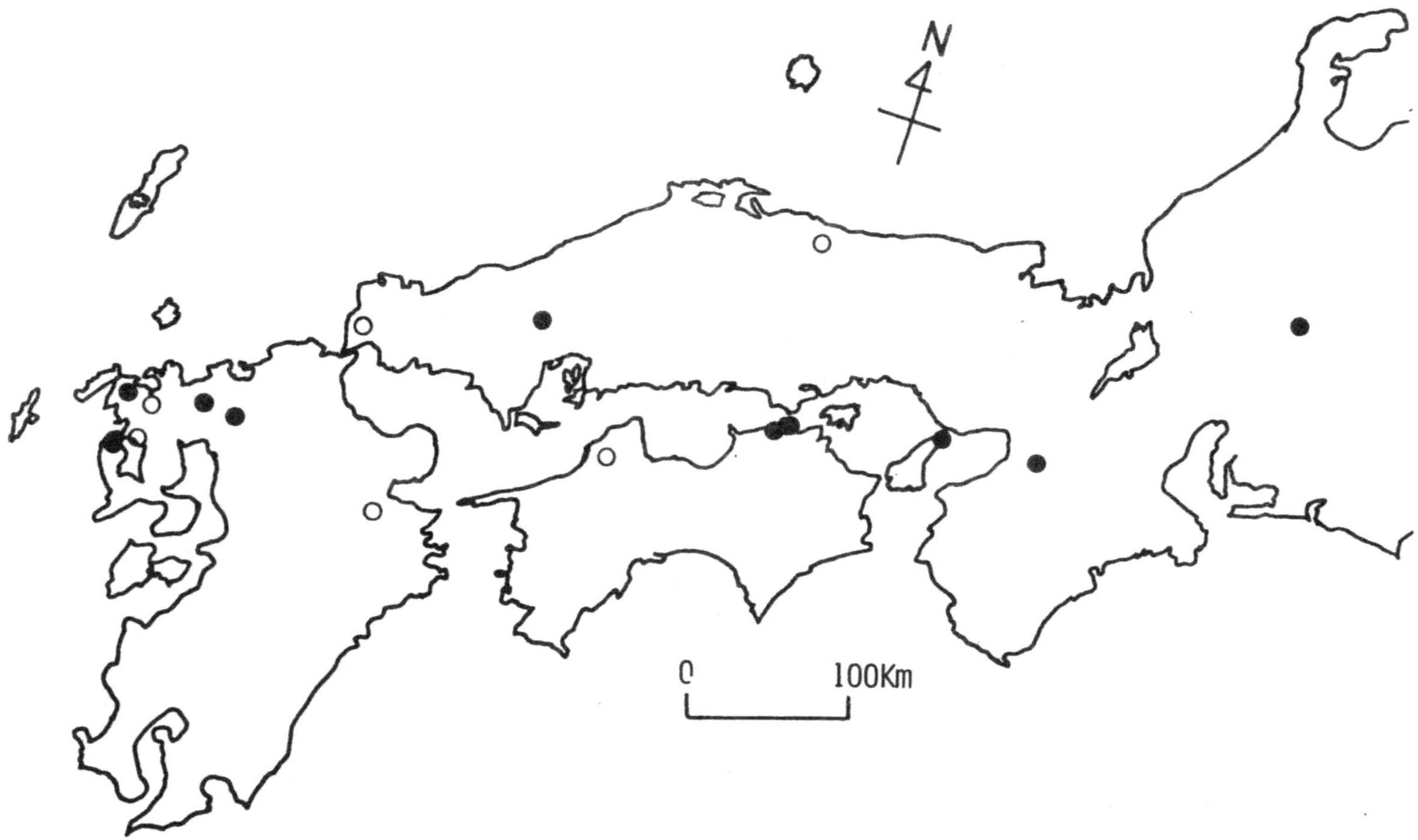


Figure 8.2    Geographical sources of sanukite.

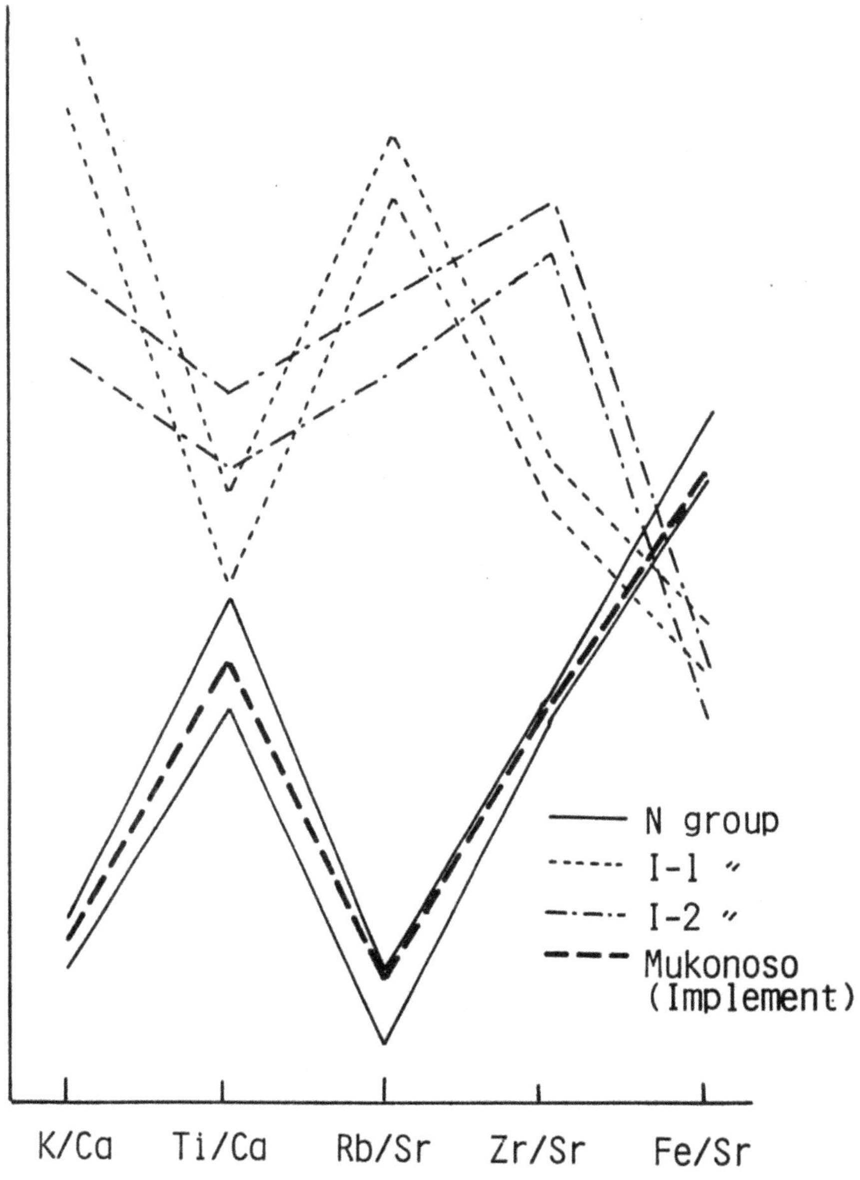


Figure 8.3 Peak-ratio patterns of sanukite stone taken from three different geographical sources (broken line) and an archaeological implement of sanukite (solid line) obtained in an archaeological site.

analyze 50 sanukite stones from every source

↓

average values of element-ratios and
standard deviations

↓

calculate variance-covariance matrices for
every source

↓

analyze an archaeological sample

↓

calculate Mahalanobis' distances from every
source

↓

F-test
(discriminant analysis)

Figure 8.4 Statistical testing procedure.

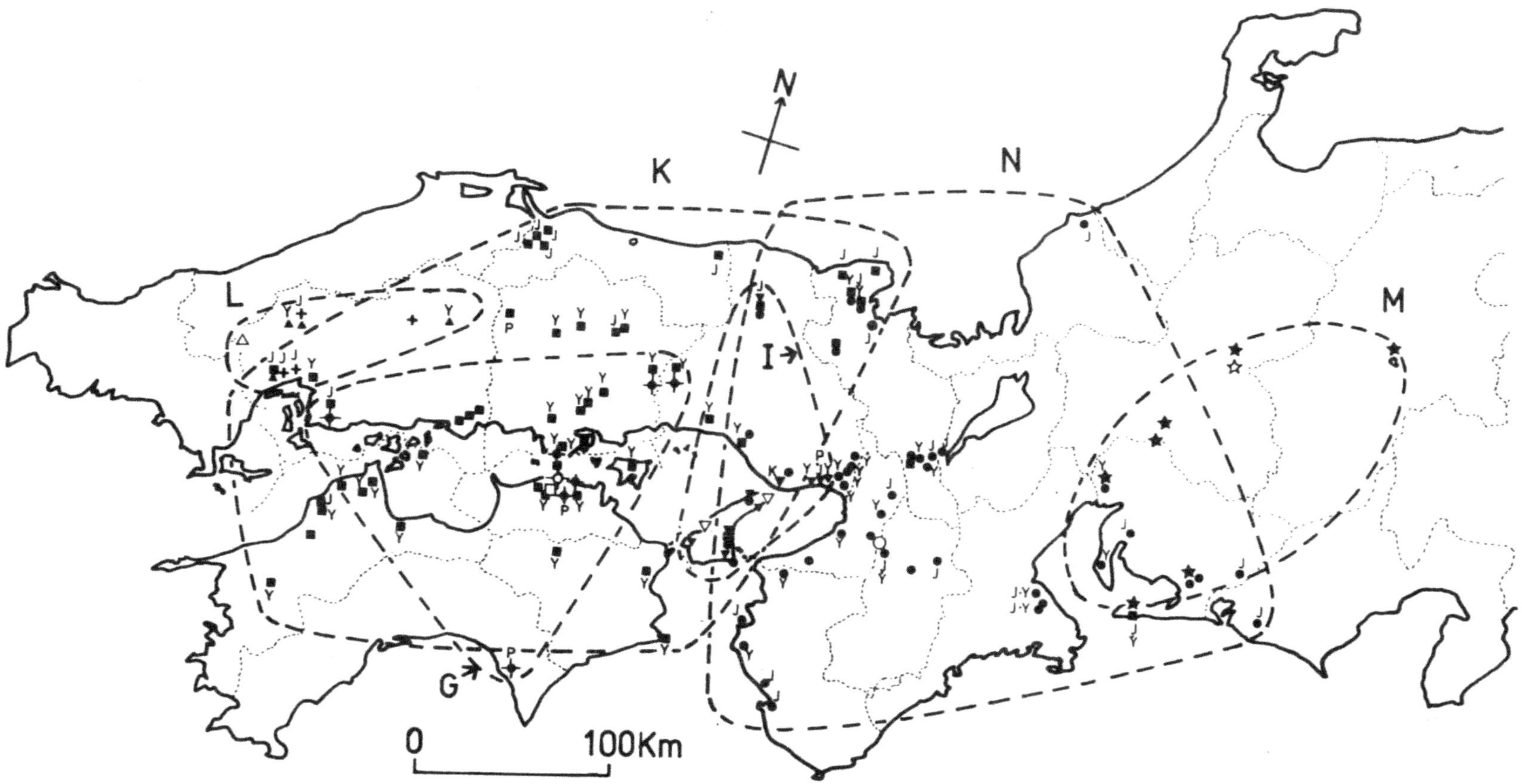


Figure 8.5 Provenances of sanukite implements as determined by X-ray fluorescence analysis. Filled marks show archaeological sites with sanukite implements from geographical sources shown by open marks of the same shapes. Areas of utilization of sanukite stones from each source are shown by broken lines.

○ : Nijohsan source (N), □ : Kanayama source (K),
-◇- : Goshikidai source (G), ▽ : Iwaya source (I),
△ : Kammuriyama source (L), ☆ : Gero source (M).

Many blocks of sanukite were taken from each geographical source and analysed by X-ray fluorescence. In some sources, peak-ratio values of different samples were highly consistent with each other indicating small intra-source variation; but in other sources, samples were classified into more than two sub-groups by their peak-ratio values. A minimum of 30 samples was used for reconstructing source groups.

Because of intra-source variation, the peak-ratio pattern for a source must be shown by a pair of lines whose width is equal to twice the standard deviation of the source group. In Figure 8.3, the patterns of several source groups are shown by pairs of broken lines. A thick broken line, which is shown as "Mukonoso" in the figure, is an implement excavated in the Tano archaeological site. Clearly, this pattern belongs to the N-group. Therefore, this sanukite implement was made from a stone taken from the Nijohzan locality.

Although this pattern method is very good for visual demonstration, a conclusion obtained by this method is only qualitative. In order to obtain a quantitative conclusion, we have to carry out a discriminant analysis on a sample.

There are several different statistical methods which have been applied to the problem of allocation of archaeological objects. We believe that the best method is the Hotteling's T square test which was used to allocate obsidian implements in New Zealand by Ward of Otago University (Ward, 1974b ; Figure 8.4).

Although fresh surfaces of sanukite stones look completely black, the archaeological samples of sanukite implements look slightly grey. This is because the surface has weathered after the implements were buried. The thickness of the weathered layer is several tens of microns or more depending on the elemental composition of the sanukite implement and the soil condition of the archaeological site. In order to reduce the error in element analysis caused by weathering, the surface layer is removed by grinding or chipping. Thus, the fluorescence X-rays are counted from a fresh surface.

Up to this time, we have analysed more than 500 sanukite implements taken from 170 archaeological sites. The result is shown in Figure 8.5. Each filled mark shows an archaeological site from which sanukite implements were obtained and analysed. Geographical sources of these implements are shown by open marks of the same shape. The area of transportation of sanukite stones from each source is shown with a series of dashed lines. The K area is very wide, covering the whole part of Shikoku island, and extends to the north coast of Western Japan. This was not expected by archaeologists before our result was obtained. The geographical source, K, for this area is marked as an open rectangle on the map. The sanukite stone was transported more than 200 km beyond Setonaikai, the inland sea. We found many archaeological sites where implements were made from sanukite stones of two or three different geographical sources. Such a fact has already been found by Ward (1974a) in the case of obsidian implements in New Zealand, and also by Suzuki (1973) in the case of obsidian implements in Eastern Japan. It therefore appears that Neolithic people generally obtained their stone from various geographical sources, and that they were enterprising in their search for good materials for making implements.

ACKNOWLEDGEMENTS

The authors are grateful to members of the departments of Archaeology of several universities and many archaeologists belonging to the committees of education of different prefectures who kindly offered us archaeological samples.

REFERENCES

Aspinall, A. and Feather, S. W., 1972. Neutron activation analysis of prehistoric flint mines. Archaeometry 14(1): 41-53.

De Bruin, M., Korthoven, P. J. M., Bakels, C. C. and Groen, F. C. A., 1972. The use of non-destructive activation analysis and pattern recognition in the study of flint artefacts. Archaeometry 14(1): 55-63.

Gordus, A. A., Fink, W. C., Hill, M. E., Purdy, J. C. and Wilcox, T. R., 1967. Indentification of the geologic origins of archaeological artifacts: an automated method of Na and Mn neutron activation analysis. Archaeometry 10: 87-96.

Higashimura, T. and Warashina, T., 1975. Sourcing of sanukite stone implements by X-ray fluorescence analysis. Journal of Archaeological Science 2: 169-178.

Nelson, D. E., d'Auria, J. M. and Bennett, R. B., 1975. Characterization of Pacific north-west coast obsidian by X-ray fluorescence analysis. Archaeometry 17: 85-97.

Smith, I. E. M., Ward, G. K. and Ambrose, W. R., 1977. Geographic distribution and the characterization of volcanic glasses in Oceania. Archaeology and Physical Anthropology in Oceania 12(3): 173-201.

Suzuki, M., 1973. Chronology of prehistoric human activity in Kanto, Japan. Journal of the Faculty of Science, the University of Tokyo Section 5(4): 241-318, 395-469.

Warashina, T., Kamaki, Y. and Higashimura, T., 1978. Sourcing of sanukite implements by X-ray fluorescence analysis II. Journal of Archaeological Science 5: 283-291.

Ward, G. K., 1974a. A paradigm for sourcing New Zealand archaeological obsidians. Journal of the Royal Society of New Zealand 4(1): 47-62.

Ward, G. K., 1974b. A systematic approach to the definition of sources of raw material. Archaeometry 16: 41-53.

# 9. STATISTICAL ANALYSIS OF MULTI-ELEMENT DATA FROM PREHISTORIC SOURCES OF RAW MATERIAL--SOME GEOCHEMICAL CONSIDERATIONS

G. M. Kovnurko, V. I. Mishin and G. I. Suslov

## INTRODUCTION

The problem of sources of raw material in archaeology is many faceted. Sampling, the study of the properties and composition of various rocks, and the processing of observed results are inter-related operations. One of the latest surveys of this field in archaeology is by Shotton and Hendry (1979).

The chemical analysis of archaeological material produces data of a statistical character, and if petro-archaeological correlations are to be made, it is essential that these are accompanied by a suitable reliability estimate. To this end, not only should a large number of objects be analysed, but a large number of properties should be examined also. A wide variety of techniques have been applied to this problem, including the presence or absence of certain features, ternary plots, cluster analysis, discriminant function analysis, and factor analysis.

Ward, in his study of New Zealand obsidians (1974) reviewed previously used statistical methods, and successfully applied the multivariate Mahalonobis generalised distance statistic to classify artefacts to sources in New Zealand.

It should be noted that the geochemical character of different rocks varies considerably, and the statistical approach such as Ward's would have to be individually worked out to suit each rock type. In our view, the important feature which should be taken into account is the pattern of distribution of trace elements—in particular, the range of variation of each, and the inter-correlation of one with another. The whole question of geochemical correlation is assuming a more important role in such studies, because as analytical techniques have been refined, the sample sizes have diminished. This means that the possibility of sample bias is now a more serious problem than ever before.

The present paper is concerned with this question for obsidian and flint —two stone types widely used by prehistoric man.

## OBSIDIAN

This material is a glassy substance, rich in silica, and of volcanic origin. When subjected to petrographic and geochemical analysis, obsidian reveals its heterogeneous character. Its properties and composition vary from one volcano to another, and this fact can be put to good use in sourcing studies. More to the point, as Bowman et al. (1973) have shown, there can be changes in concentration of some trace elements within the confines of a single lava flow. The lava flow studied in Northern California consisted of a dacite layer overlain by obsidian. The concentrations of Fe, Co, Mn, and Sc were found to decrease as samples were taken further up the section of the outcrop, and concentrations of Th, U, and Cs were found to increase. Such regularity is typical for magmatic rocks—there is a progressive increase in the lytophile elements relative to siderophile ones with increasing distance up the section of the flow.

This theory is completely borne out by the investigations by Leonova (1974) of acid volcanic rocks in Kamchatka. In order to discover the pattern of distribution of trace elements in volcanic rocks of the Uzonsko-Geizernaia Depression, and of the Bolshoi Semjachic structure, factor analysis was carried out. The results are illustrated in Figure 9.1 and 9.2. This analysis convincingly demonstrates that within each volcanic structure there is an increase in the concentration of Li and Rb at the expense of Co and V. These four elements form the first factor structure, which has an overall weight of 28.13%. These elements cannot be used to separate the two suites of rocks. However, the two types of rocks differ in the concentrations of elements which form the structure of the second factor (weight 23.94%), especially Th and U.

Thus, a peculiar feature of factor analysis is its ability to reveal, amongst the large number of elements present, those particular ones which specify geochemical specialization in different volcanoes, and consequently which can be used to compare geological and archaeological materials.

In volcanic rocks, including obsidian, there is some heterogeneity arising from partial recrystalization of the magmatic melt. This is characterized by microlite and porphyritic dissemination of alkaline feldspars and cristobalite porphirocrysts. The size of the dissemination greatly varies: in some cases they can only be seen with an electron microscope, while in other cases they can be seen in normal petrographic thin sections (Nasedkin, 1975).

In obsidian samples of small size, there could be considerable variation from one to another of K, Na, Ca geochemically consanguineous Rb, Ba, Sr, and some other elements. These are part of the high temperature feldspars, and are highly correlated. Factor analysis can establish how important these elements are when comparing obsidian artefacts with geological sources, and how necessary it may be to use petrographic techniques to check on dissemination. To this end, factor analysis was carried out using the element data published by Cann and Renfrew (1964), and by Stevenson et al. (1971, NB: in some cases they give only average obsidian compositions for N analyses. In these cases the factor analysis was carried out on N samples with these average values). Figures 9.3 and 9.4 show the results for the first two factors in each case.

The first two factor diagrams show considerable similarity in the arrangement of correlated chemical elements. For instance, the upper right square contains the alkali-earth elements, and alkaline elements are to the left near the axis of the second factor.

Thus, cations forming feldspars play the most important part in geochemical classification of obsidians. Petrographic study of samples is necessary to determine the quantity and size of feldspar dissemination in volcanic rocks. All samples showing anomalous concentrations of alkalis or alkali-earths should be excluded from further consideration.

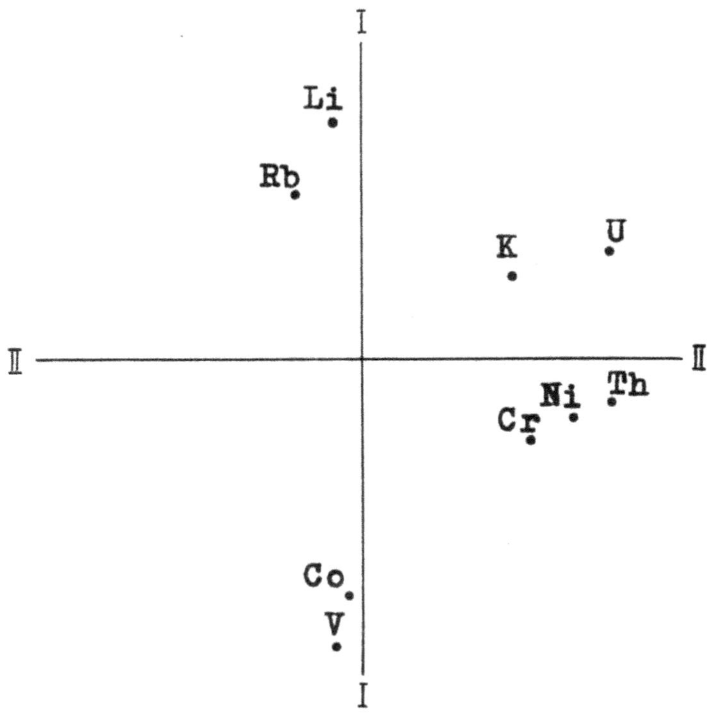


Figure 9.1 The first two factor loadings from the analysis of Kamchatka acid volcanic rocks (after Leonova, 1974).

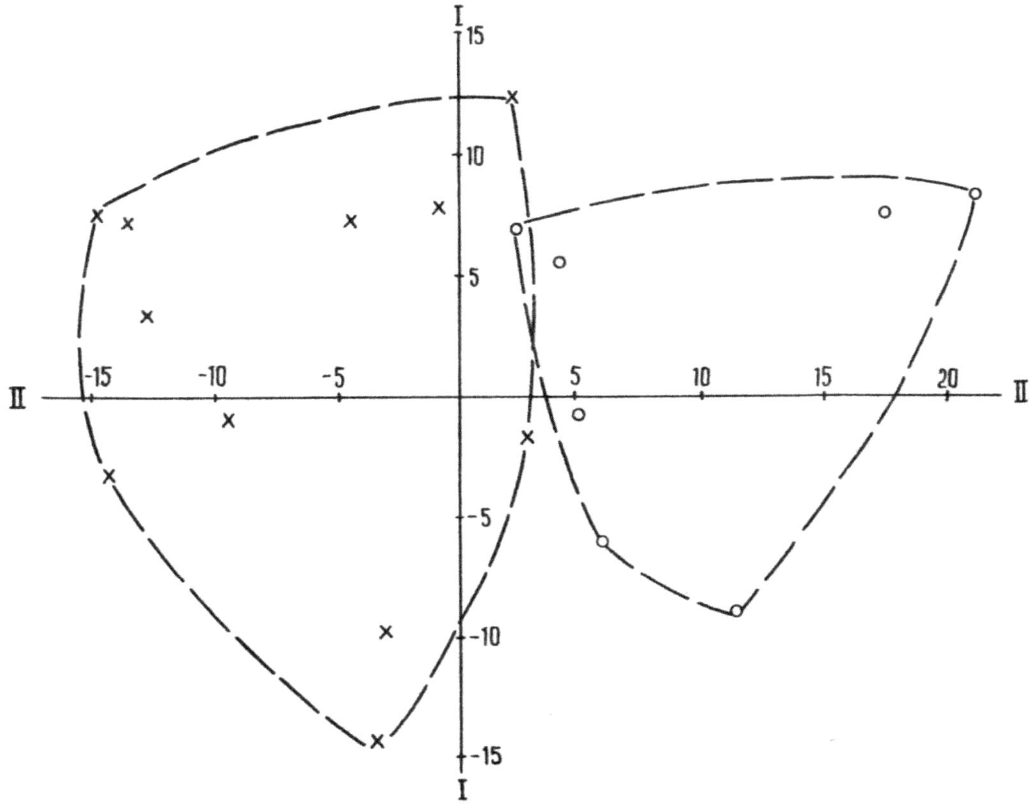


Figure 9.2 The first two factor loadings from the analysis of Kamchatka acid volcanic rocks (after Leonova, 1974). x Bolshoi Semjachic structure, o Uzonsko-Geizernaia Depression.

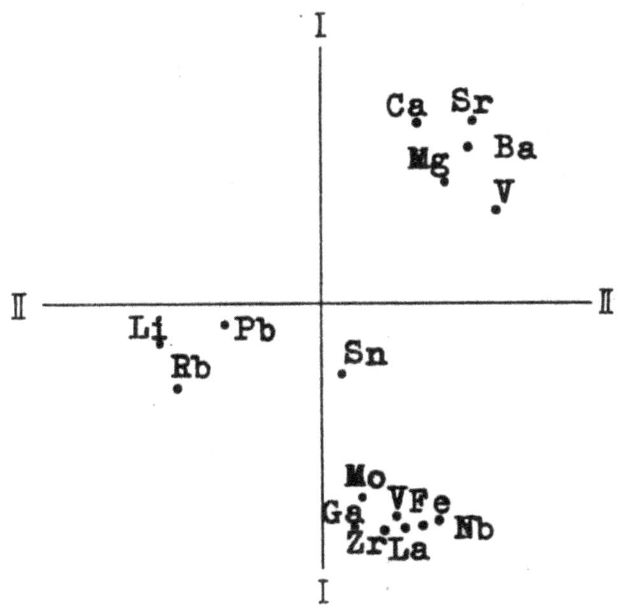


Figure 9.3 The first two factor loadings from the analysis of obsidian data given by Cann and Renfrew (1964).

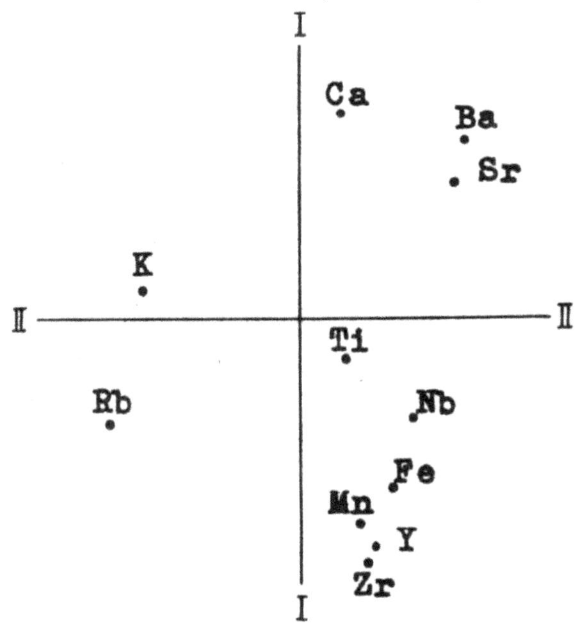


Figure 9.4 The first two factor loadings from the analysis of obsidian data given by Stevenson _et al_. (1971).

The above examples testify that factor analysis of obsidian chemical composition data has some unique advantages over other types of statistical analysis used in sourcing studies. With the aid of this method it should be possible to draw up a list of elements which are most effective in characterizing the raw material. The structural analysis of the principle factors shows that the distribution of alkaline and alkali-earth elements in obsidians might depend on the petrographic pecularities of this kind of volcanic rock—the use of these elements therefore in obsidian sourcing studies should only be made with great caution.

There are many discrete obsidian sources on the earth's surface, and it is very worthwhile to establish a catalogue of these, using multi-element analysis.

FLINT

In archaeological literature, far too much attention is given to obsidian at the expense of flint. Part of the reason for this is the peculiarity of geological and geochemical properties of flint, which are insufficiently understood at present. In our view (Kovnurko, 1970; 1973), flint is formed in a mass of carbonate silt through diagenesis. Concretions of silica form as a result of certain chemical reactions taking place when the pH value of porous solutions of carbonate silt decreases due to microbiological decomposition of the organic component. The low percentage of impurities in flint is a result of the peculiar conditions in its formation—spectral analysis shows some flints to be much purer thanrock crystals.

The following pecularities should be taken into account when carrying out petro-archaeological studies of flints.

* Flint concretions in interlayers are distributed almost continuously. Some layers may stretch for tens or even hundreds of kilometres.
* The unique element composition and other properties of flint within any one interlayer are very uniform, and are not related to those characteristics of other layers either above or below.
* The spacing of flint interlayers in a stratigraphic section varies considerably. Analytical techniques can establish whether a flint sample belongs to a particular interlayer. However, the fact that these highly uniform layers extend over great distances makes it very difficult to determine the geographic source of any piece of flint. This problem is partly overcome by the fact that prehistoric man, in his mining operations, made definite choices of flint amongst the enormous variety available. For this reason, the study of flint implements from large prehistoric mines, pits and factories is of great importance to establish both the petrographic and geochemical characteristics of the flint actually used, and where it came from.

Data from spectrographic and atomic absorption analysis have been used by Sieveking et al. (1970; 1972) for classifying flint from seven West-European pits and factories. They analysed 20 samples from each archaeological site, and determined the concentration of 8 elements. With the aid of discriminant analysis, the authors established a percentage impurity for each type of flint.

Reference collections of flint from different areas are of central importance for the archaeologist studying the trade routes of flint implements. In order to find out the geochemical pecularities of flint, we have studied the results of spectral analyses carried out on 975 samples by K. Kynchev of Bulgaria. Our intention has been to establish the distribution patterns of Al, Ti, Ca, Mg, Fe, Mn, Na, Cu, Zn and Cr within the limits of separate layers. These patterns will allow us to differentiate between material of different interlayers.

The first stage of this study has been to apply factor analysis (using the method of principle components) to the spectral data. A few weak positive correlations between some elements have been determined, and this confirms these as forms of impurity in the flint. The greatest number of elements are correlated with Al (at the 99% significance level), and this fact then confirm the hypothesis suggested by Sieveking et al. (1972). According to this hypothesis, flint consists of clay minerals along with carbonates, oxides, and structureless impurities.

Since only weak correlations were found, the eigenvalues for each factor extracted are not greatly different, and the substitution of element concentrations for factors does not result in a decrease in the number of variables. This result made it necessary to adopt a second line of approach. A non-hierarchical classification of the multi-element data was attempted, by division of the total assemblage into groups by means of a minimum branching dendrogram. The 975 flint samples were thus classified into 19 groups. Discriminant analysis was used to assess the relative power of different variables to distinguish groups, and to test the significance of group separation. The 19 groups discerned, using the 10 elements enumerated, are only tentative at this stage. It appears that there is a maintenance of element concentrations within separate concretions as well as within separate interlayers of flint, and the changes are fairly continuous and overlapping. Cataloguing the composition of flint is further complicated by the presence of reprecipitation flint in Quaternary sediments.

## CONCLUSION

In the statistical analysis of multi-element data for the purposes of distinguishing between sources of material, the choice of technique should largely depend on the inter-element correlation matrix. In cases of high correlation, factor analysis is able to establish the most informative elements, which can then be used to characterize the original sources. In cases of a structureless correlation matrix, cluster analysis, accompanied by discriminant analysis, may be a more effective alternative.

## REFERENCES


Bowman, H. R., Asaro, F. and Perlman, I., 1973. Composition variation in obsidian sources and the archaeological implications. Archaeometry 15(1): 123-127.

Cann, J. R. and Renfrew, C., 1964. The characterization of obsidian and its application to the Mediterranean region. Proceedings of the Prehistoric Society 30(9): 111-133.

Kovnurko, G. M., 1970. Calcite, alpha-cristobalite and quartz in zonal concretions of flint. Proceedings of Academy of Sciences, USSR 195(4): 932-934.

Kovnurko, G. M., 1973. Concretions of flint of a zonal structure. Leningrad University Bulletin, Geological and Geographical Series 24: 68-72.

Leonova, L. L., 1974. Geochemical characteristics of basic and acid volcanic rocks of the Uzonsko-Geizernaia Depression. pp. 50-60 In: Vulcanism, the hydrothermal process and ore-formation. Nedra, Moscow.

Nasedkin, V. V., 1975. The petrogenesis of acid volcanic rocks. Nauka, Moscow.

Shotton, F. W. and Hendry, G. L., 1979. The developing field of petrology in archaeology. Journal of Archaeological Science 6(1): 75-84.

Sieveking, G. de G., Craddock, P. T., Hughes, M. J., Bush, P. and Ferguson, J., 1970. Characterization of prehistoric flint mine products. Nature 228: 251-254.

Sieveking, G. de G., Bush, P., Ferguson, J., Craddock, P. T. and Hughes, M. J., 1972. Prehistoric flint mines and their identification as sources of raw material. Archaeometry 14(2): 151-176.

Stevenson, D. P., Stross, F. H. and Heizer, R. F., 1971. An evaluation of X-ray fluorescence analysis as a method for correlating obsidian artifacts with source location. Archaeometry 13(1): 17-25.

Ward, G. K., 1974. A systematic approach to the definition of sources of raw material. Archaeometry 16: 41-53.

КОВНУРКО Г.М.
(Ленинградский Государственный
Университет им.А.А.Жданова)
МИШИН В.И., СУСЛОВ Г.И.
(Северо-Западное Территориальное
Геологическое Управление)

# СТАТИСТИЧЕСКАЯ ОБРАБОТКА РЕЗУЛЬТАТОВ МНОГОЭЛЕМЕНТНОГО АНАЛИЗА ГОРНЫХ ПОРОД В ЦЕЛЯХ ПОИСКА КОРЕННЫХ ЗАЛЕЖЕЙ КАМЕННОГО СЫРЬЯ (ГЕОХИМИЧЕСКИЙ АСПЕКТ)

Проблема поиска коренных залежей каменного сырья в археологии многоаспектна. Отбор проб для анализа, изучение свойств, состава каменного материала, обработка результатов наблюдений являются взаимоопределяющими друг друга операциями. Один из последних обзоров достигнутых в этой области археологии успехов приведен в работе **Shotton and Hendry** (1979).

Статистический характер наблюдений при аналитическом изучении археологического материала является очевидным и обусловлен необходимостью оценки надежности петроархеологических сопоставлений. При этом под статистическим характером наблюдений подразумевается не только количественное изучение многочисленных объектов, но и определение у каждого объекта возможно большего числа признаков, свойств. При обработке результатов наблюдений в петроархеологических исследованиях наиболее широко используются бинарные и вариационные диаграммы, кластерный, дискриминантный и факторный анализы. Работа **Word** (1974) посвящена исследованию обсидианов Новой Зеландии. Автор приводит сведения о применявшихся ранее приемах статистического анализа многомерных данных и с успехом использует обобщенное расстояние Махалонобиса для классификации обсидианов из естественных залежей Новой Зеландии и сопоставления с ними материала из археологических памятников.

Следует отметить, что геохимические особенности различных горных пород требуют индивидуального аналитического подхода и специфических методов статистической обработки результатов анализов. Под геохимическими особенностями горных пород, которые важно учитывать при поисках источников каменного сырья, мы понимаем характер распределения малых элементов в горных породах, в первую очередь их взаимосвязи и масштабы колебания концентраций. Вопросы геохимии стали актуальными в связи с тем, что по мере совершенствования аналитической техники размеры анализируемых проб уменьшаются, т.е. пробы становятся все менее представительными.
В докладе затрагиваются геохимический и математический аспекты проблемы поиска коренных залежей обсидиана и кремня - горных пород, наиболее широко использованных для изготовления орудий труда.

Обсидиан.

Стекловатая вулканическая горная порода с высоким содержанием двуокиси кремния при петрографическом и геохимическом изучении обнаруживает гетерогенность.
С точки зрения поиска источников каменного сырья наиболее важным являются различия в составе, свойствах между обсидианами отдельных вулканов. Но в пределах одного лавового потока возможно изменение концентраций некоторых редких элементов, как показали **Bowman, Asaro and Perlman** (1973). Изученный исследователями в Северной Калифорнии лавовый поток был сложен слоем дацита, перекрывавшегося обсидианом, причем снизу вверх по разрезу уменьшались концентрации **Fe, Co, Mn, Sc** и увеличивались содержания **Th, U, Cs.** Такая закономерность является типичной для магматических образований: увеличение снизу вверх по разрезу содержания литофильных элементов, относительно сидерофильных.

Подтверждением этому являются результаты изучения кислых вулканитов Камчатки (Леонова Л.Л., 1974). Для выяснения закономерностей в распределении редких элементов в вулканических горных породах Узоно-Гейзерной депрессии и структуры Большой Семячик был проведен факторный анализ результатов изучения редкоэлементного состава пород методом главных компонент, результаты которого представлены на рис.1 и 2. Факторный анализ убедительно показывает, что в пределах каждой вулканогенной структуры происходит увеличение концентраций Li и Rb за счет Co и V. Эти элементы входят в структуру первого фактора, вес которого составляет 28,13%. Они не могут быть использованы для разделения вулканитов Узоно-Гейзерной депрессии от вулкана Большой Семячик. Горные породы двух вулканов хорошо различаются по элементам, вошедшим в структуру второго фактора (вес 23,94%), в первую очередь по U и Th.

Таким образом, существенной особенностью факторного анализа является возможность выявления среди множества химических элементов тех, которые обуславливают геохимическую специализацию горных пород разных вулканов и которые должны быть использованы при сопоставлении геологического и археологического материала.

В вулканических породах, в том числе в обсидианах наблюдается гетерогенность, связанная с частичной раскристаллизацией магматического расплава и выражающаяся в появлении микролитов, порфировых вкрапленников щелочных полевых шпатов и кристобалита. Размеры вкрапленников колеблются в широких пределах: в одних случаях они различимы лишь под электронным микроскопом, в других - видны в петрографических шлифах (Наседкин, 1975). При отборе проб обсидиана небольшого объема возможны значительные вариации в содержании K, Na, Ca и близких к ним геохимически Rb, Ba, Sr, Cs и др.элементов, входящих в структуру высокотемпературных полевых шпатов и связанных между собой прочной корреляционной связью. Факторный анализ позволяет выяснить, насколько важны эти элементы при сопоставлении обсидиана из коренных залежей с материалом артефактов, т.е. насколько необходим петрографический контроль. С этой целью была проведена статистическая обработка состава обсидианов, анализы которых приведены в работах Cann and Renfrew (1964), Stevenson, Stross and Heizer[x] (1971). На рис.3 и 4 изображены значения факторных нагрузок первого и второго факторов.

Изучение диаграмм показывает большое сходство во взаимном расположении многих химических элементов. В верхнем правом квадрате сосредоточены щелочноземельные элементы, а слева близ оси второго порядка - щелочные. Таким образом, катионы, слагающие полевые шпаты, играют важную роль при геохимической классификации обсидианов и петрографическое изучение образцов необходимо для выяснения количества и размера вкрапленников полевых шпатов в вулканической породе. Пробы, выделявшиеся аномальными концентрациями щелочей и щелочных земель должны быть исключены из дальнейшей обработки.

Приведенные примеры свидетельствуют о том, что факторный анализ редкоэлементного состава обсидианов имеет ряд преимуществ перед другими методами статистической обработки подобных петроархеологических наблюдений. С помощью этого метода можно установить список химических элементов, использование которых при поисках источников каменного сырья наиболее эффективно. Анализ структуры главных факторов показывает, что ха-

---

[x] В работе Stevenson, Stross and Heizer (1971) приведены в некоторых случаях средние составы обсидиана из n анализов. При введении исходных данных в ЭВМ в этом случае средний состав повторялся n раз.

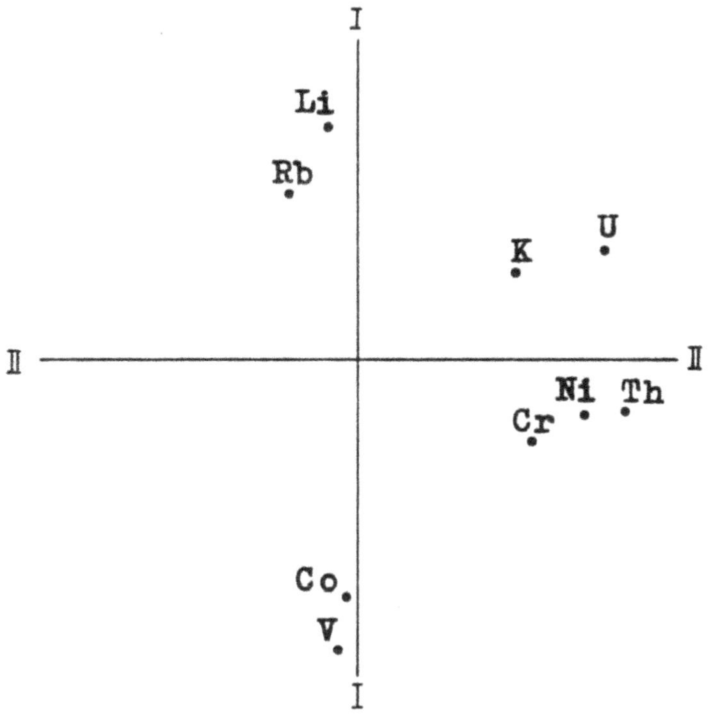


Рис.I. Факторные нагрузки I и П факторов по результатам изучения состава кислых эффузивов Камчатки (Леонова, 1974)

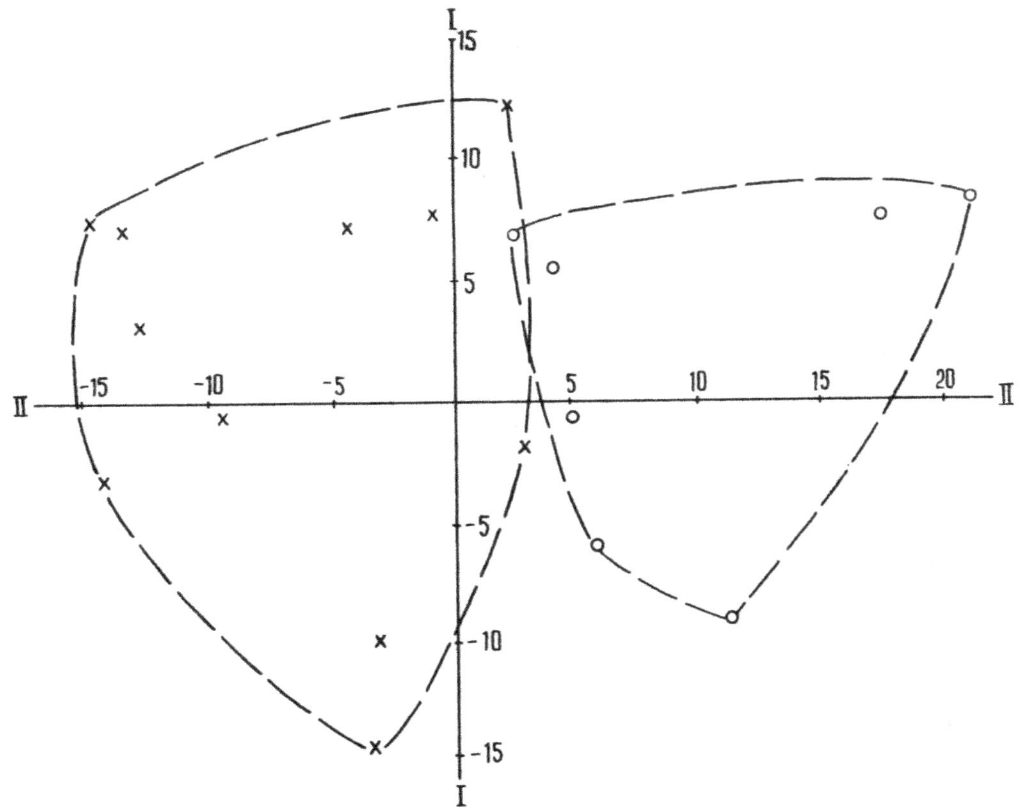


Рис.2. Значения I и П факторов, вычисленные для проб кислых эффузивов Камчатки

× Большой Семячик
o Узоно-Гейзерная депрессия

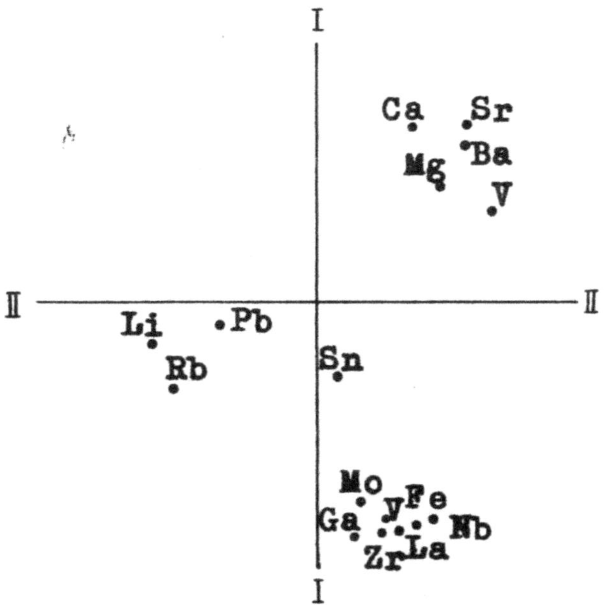


Рис.3. Факторные нагрузки I и П факторов по результатам состава обсидианов
**(Cann and Renfrew, 1964)**

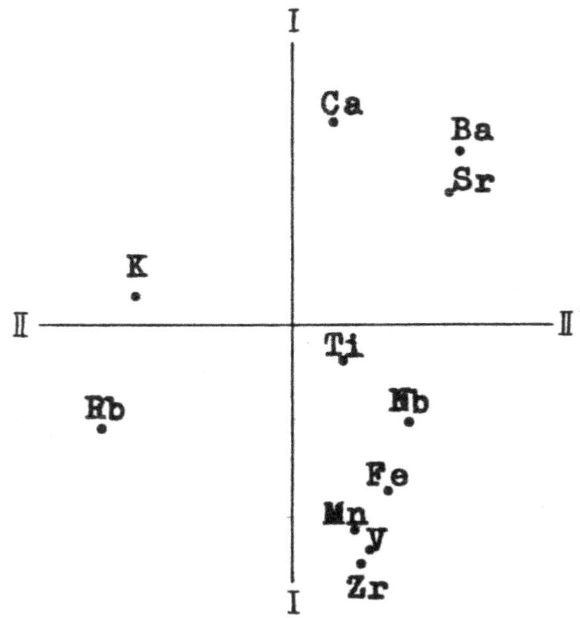


Рис.4. Факторные нагрузки I и П факторов по результатам изучения состава обсидианов
**(Stevenson, Stross and Heizer, 1971)**

рактер распределения щелочных и щелочноземельных элементов в обсидианах может быть обусловлен петрографическими особенностями этих вулканогенных пород.

Дискретное распределение обсидиана на земной поверхности благоприятствует созданию каталогов многоэлементных анализов вулканического стекла из естественных местонахождений.

Кремень.

Поиску источников кремня, послужившего материалом для артефактов, посвящено в археологической литературе значительно меньше работ, чем обсидиану. Это объясняется своеобразными геолого-геохимическими особенностями кремня, которые в настоящее время слабо изучены.

В соответствии с нашими представлениями (Ковнурко, 1970, 1973) кремень образуется в толще морского карбонатного осадка на стадии диагенеза. Кремнезем стягивается в конкреции в результате химических реакций, происходящих там, где снижена величина pH поровых растворов карбонатного ила вследствие микробиологического разложения органического вещества. Своеобразие условий образования кремня обуславливает малое содержание в нем элементов-примесей: некоторые кремни, по данным спектрального анализа чище горного хрусталя.

Следующие особенности важно учитывать при петроархеологическом исследовании кремня.

1. Конкреции кремня в пределах прослоев распределены практически непрерывно. Отдельные слои могут простираться на расстояния в десятки, иногда сотни километров.

2. В пределах одного прослоя свойства и состав кремня устойчиво выдерживаются и не зависят от свойств и состава соседних прослоев.

3. Расстояния между соседними прослоями кремня в стратиграфическом разрезе колеблются в широких пределах.

Таким образом, с помощью аналитических методов можно установить принадлежность образца кремня к прослою, который обнажен нередко в удаленных друг от друга пунктах и это значительно усложняет поиск источников каменного сырья. В связи с этим большое значение приобретает исследование каменного материала крупных доисторических разработок кремня, шахт и мастерских по обработке. Обстоятельством, благоприятствующим таким петроархеологическим исследованиям, является то, что доисторический человек при добыче кремня обычно останавливал свой выбор на какой-то определенной разновидности кремня среди многочисленных его прослоев.

Результаты спектрального и атомноабсорбционного анализов использовали Sieveking at al. (1970, 1972) для классификации кремня из семи западноевропейских мастерских и шахт. Каждый археологический памятник был представлен выборкой из 20 образцов, в которых определялись концентрации восьми химических элементов. С помощью дискриминантного анализа удалось установить состав элементов-примесей в кремне каждого из семи памятников.

Значительный интерес для археологов, изучающих пути обмена торговли кремневым инвентарем, представляют эталонные коллекции кремня из естественных коренных местонахождений. Для выяснения геохимических особенностей кремня в наше распоряжение были предоставлены К.Кынчевым результаты спектрального анализа 975 проб из Болгарии. Целью работы явилось выявление законов распределения Al, Ti, Ca, Mg, Fe, Mn, Na, Cu, Zn и Cr в пределах отдельных слоев и позволяющих различать материал отдельных прослоев. Результаты спектрального анализа горных пород первоначально изучались с помощью факторного анализа (метод главных компонент). На первом этапе работы были установлены немногочисленные слабые положительные корреляционные связи между некоторыми элементами. Это свидетельствует о разнообразных формах нахождения примесей в кремне. Наибольшим числом значимых на 99% уровне связей отличается Al и это наблюдение можно расценивать как подтверждение предположения, высказанного в работе Sieveking at al. (1972), о том, что наряду с карбонатами, окислами, бесструктурными примесями в кремне присутствуют глинистые минералы. Незначительное число слабых связей в корреляционной матрице является причиной того, что главные факторы не различаются по весу между собой, а замена концентраций отдельных элементов значениями факторов не приводит к сокращению числа переменных.

Эти обстоятельства привели к необходимости использования для группировки проб неиерархической классификации многомерных данных (разбиение совокупности на группы при помощи дерева минимального покрытия). 975 проб кремня были разделены на 19 групп, а проверка качества разделения и оценка информативности переменных проводилась с помощью обобщенного дискриминантного анализа. Разделение на группы в соответствии с концентрациями указанных выше 10 элементов является условным, поскольку содержание этих элементов в пределах отдельных конкреций и в пределах прослоев изменяется постепенно и число групп определяется параметрами, вводимыми в компьютер. Составление каталога естественного кремня усложняется также присутствием переотложенного кремня в четвертичных отложениях.

Таким образом, при статистической обработке результатов многоэлементного анализа горных пород в целях поиска коренных залежей каменного сырья обнаружение многочисленных прочных корреляционных связей между элементами способствует последующему использованию метода главных компонент факторного анализа, с помощью которого представляется возможность выявить наиболее информативные химические элементы. В тех случаях, когда корреляционная матрица бесструктурна, для классификации многоэлементных анализов горных пород могут быть использованы приемы дискриминантного и кластерного анализов.

ЛИТЕРАТУРА REFERENCES


I. Ковнурко Г.М. 1970. Кальцит, $\alpha$ -кристобалит и кварц в зональных конкрециях кремня. Доклады АН СССР, т.195, № 4, с.с. 932-934.

2. Ковнурко Г.М. 1973. Конкреции кремня зонального строения. Вестник Ленинградского Университета, серия геолог., географ., № 24, с.с. 68-72.

3. Леонова Л.Л. 1974. Геохимическая характеристика основных и кислых вулканитов Узонско-Гейзерной депрессии. В кн.„Вулканизм, гидротермальный процесс и рудообразование". М., Недра, с.с. 50-60.

4. Наседкин В.В. 1975. Петрогенезис кислых вулканитов. М., Наука.

5. Bowman H.R., Asaro F., Perlman. 1973. Composition variation in obsidian sources and archaeological implications. Archaeometry, v.15, N 1, p.p. 123-127.

6. Cann J.R., Renfrew C. 1964. The characterization of obsidian and its application to the Mediterranien region. Proceedings of the prehistoric society, v.XXX, N 9, p.p.111-133.

7. Shotton F.W., Hendry G.L. 1979. The developing fild of petrology in archaeology. J.Archaeol. Sci. v.6, N 1, p.p.75-84.

8. Sieveking G. de G., Craddock P.T., Hughes M.J., Bush P., Ferguson J. 1970. Characterization of prehistoric flint mine products. Nature, v.228, p.p.251-254.

9. Sieveking G. de G., Bush P., Ferguson J., Craddock P.T., Hughes M.J. 1972. Prehistoric flint mines and their identification as sources of raw material. Archaeometry, v.14, N 2, p.p.151-176.

10. Stevenson D.P., Stross F.H., Heizer R.F. 1971. An evalution of X-ray fluorescence analyses as a method for correlation obsidian artifacts with source location. Archaeometry, v.13, N 1, p.p.17-26.

# 11. NEUTRON ACTIVATION ANALYSIS OF NEW ZEALAND AND OCEANIC OBSIDIANS: TOWARDS A SIMPLE SCREENING TECHNIQUE

B. Foss Leach and S. E. Warren

## INTRODUCTION

The volcanic glass obsidian was a highly prized lithic material in many parts of the world during the prehistoric period. This popularity is due to several features. Firstly, an extremely sharp edged implement may be produced by a simple chipping technique, and these tools are highly suitable as knives. Secondly, obsidian is a relatively isotropic material, and is easily worked into more complex shapes by a combination of primary and secondary flaking, and in some cases by additional grinding and polishing. Thirdly, obsidian is an attractive substance with a high lustre and pure dense colouration. It is usually dark grey or black, but shades of red and green are also fairly common. Its main disadvantage as a raw material is that it is moderately brittle, and a cutting edge will not last long.

In the south Pacific basin, obsidian was clearly an important commodity at least as early as the first Lapita pottery people, who spread through much of the Oceanic world at about 3,500 years ago. The descendants of these people ultimately reached the Hawaiian chain to the north, Easter Island to the east, and New Zealand and the Chatham Islands to the south. The final stages of this colonization process were completed by about A.D. 800. The value which these people attached to obsidian is attested by the frequency with which small flake tools are found in the vicinity of sources of the raw material. The fact that obsidian tools occur in places where there is no natural supply is clear evidence that either blocks of the raw material, or tools made from it, were transported from place to place—apparently over considerable tracts of ocean on some occasions.

It has been the earnest wish of most archaeologists who have found obsidian tools in the Pacific to be able to identify the original source of raw material. It is hoped that this will reveal something of the ocean-going activities of prehistoric people, and perhaps also the nature of regular communication patterns, such as trade and economic or ceremonial exchange networks. There has been a limited success in some areas in this quest, and a notable lack of it in others. Most of the sourcing techniques which have been investigated have relied on changes in chemical composition from one source of supply to another. Recently, some reservations have been expressed about this technique, on the grounds that the same eruption of magma may surface in different volcanoes as much as a thousand miles apart, and that obsidians produced will all bear the same signature (Shutler and Shutler, 1975:102). There has been a certain amount of debate on this point by archaeologists (see for instance Smith et al., 1977:174), but on the whole there is widespread confidence in the chemical integrity of different sources. This could be false optimism because the present study of New Zealand and Pacific obsidians shows that sources as much as 4,300 miles apart (6,900 km) can have very similar composition, and must be very carefully analysed in order to separate them (see below).

SOURCING TECHNIQUES IN NEW ZEALAND - AN EVALUATION

In New Zealand there are at least 18 sources of obsidian (see Figure 11.1), and in the last 15 years archaeologists have made a number of attempts to set up techniques to distinguish these sources so that the origin of prehistoric artefacts can be determined. To be useful to the archaeologists, the sourcing method developed should meet as many of the following criteria as possible:

* It must be capable of reliably distinguishing between all potential sources which were available to prehistoric people.
* The technique should be non-destructive so that valuable cultural property can be analysed.
* It should be rapid, inexpensive, and not too complex.
* It should be available for general use by archaeologists for large runs of samples numbering in the thousands.

Unfortunately no technique so far explored in New Zealand meets all these criteria. An evaluation of the merits of the different techniques is given in Table 11.1. It will be seen that only four of the methods have actually been tested on all sources and found to be capable of distinguishing among them. These are: wavelength dispersive XRF, thermoluminescence, neutron activation analysis, and proton induced X-ray and Gamma-ray emission. Despite this apparent success, these methods have equally serious drawbacks. They are either destructive, laborious, expensive, or the equipment is simply not available for routine application to archaeological problems.

The most definitive work on this subject was carried out by Ward in 1972. Using wavelength dispersive XRF he was able to distinguish reliably between the 18 sources in New Zealand by measuring the concentrations of Zr, Mn, Ti, Rb and Sr. Multivariate discriminant functions are needed to ascribe an artefact to a particular source. It has not been possible to apply this method to many prehistoric artefacts. For one thing, a sample size of about 2 gm must be powdered, for another the method is time consuming, and existing equipment in New Zealand is heavily committed to other scientific research.

An obvious possibility is to apply Ward's results to energy dispersive XRF, which is a non-destructive method, and the machinery is generally more available for archaeological work. However, this application has not proven to be straightforward. For one thing, although obsidian might be considered isotropic, it does exhibit diffraction effects when subjected to X-ray excitation. Consequently, some element peak intensities change markedly with shifts in the geometry of the artefact surface presented for

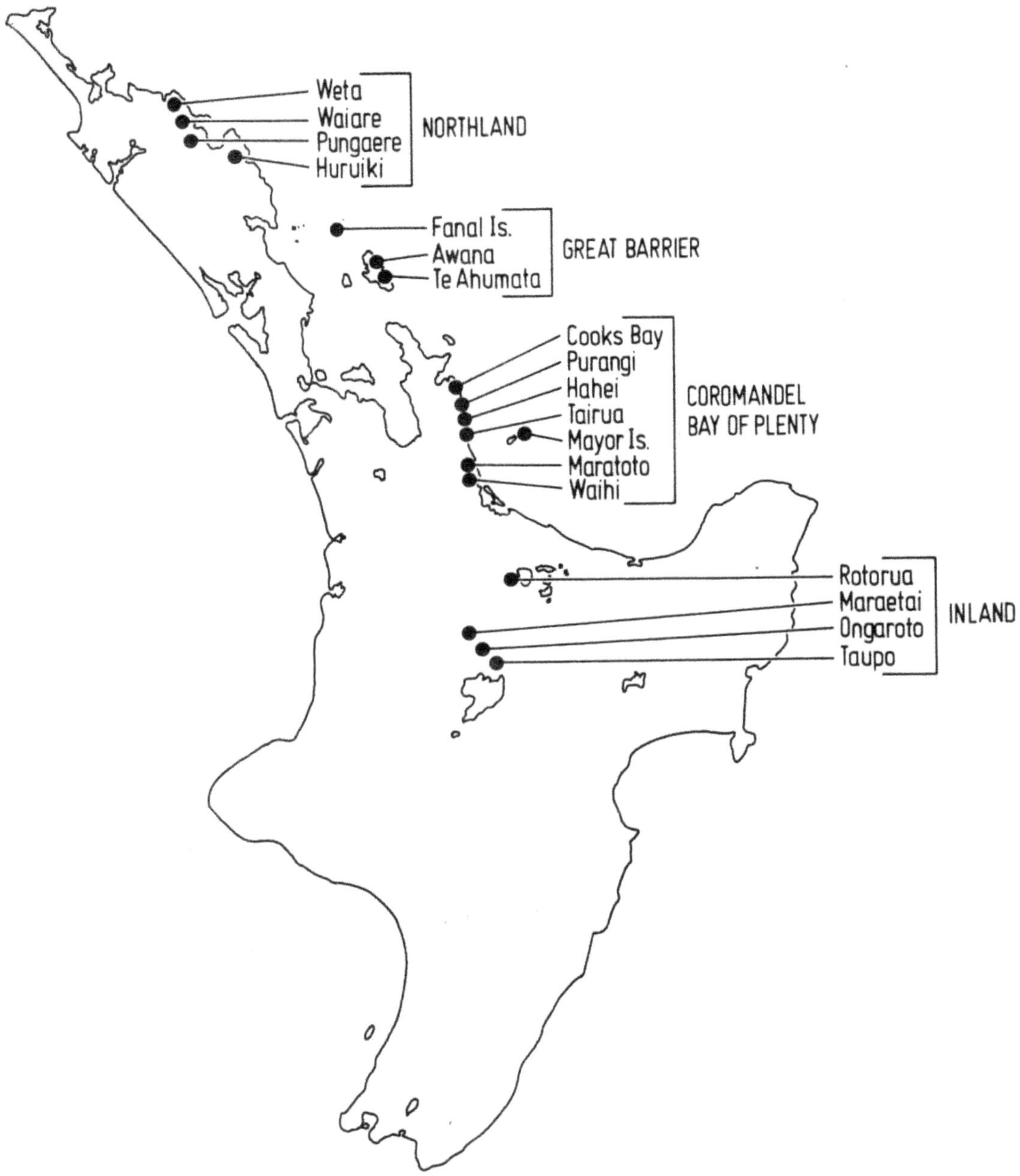


Figure 11.1 Location map of New Zealand showing obsidian sources.

TABLE 11.1 Characterization methods attempted on New Zealand obsidians - an evaluation.

| | Methods | Destructive? | Time Consuming? | Routine cost & Availability ? | Fully tested on all sources? | Source Overlaps? |
|---|---|---|---|---|---|---|
| 1 | Refractive Index[1] | - | - | Good | - | - |
| 2 | Density[2] | Good | Good | Good | - | - |
| 3 | Emission Spectroscopy[3] | - | - | - | - | Good |
| 4 | Atomic Absorption[4] | - | - | - | - | - |
| 5 | Wavelength dispersive XRF (TEFA)[5] | - | - | - | Good | Good |
| 6 | PIGME[6] | Good | Good | - | - | - |
| 7 | Thermoluminescence[7] | - | - | - | Good | Good |
| 8 | Radioactive properties[8] | Good | Good | Good | Good | * |
| 9 | Energy dispersive XRF (TEFA)[9] | Good | Good | Good | - | * |
| 10 | Energy dispersive XRF (SEFA)[10] | Good | - | Good | - | * |
| 11 | NAA[11] | - | - | - | Good | Good |
| 12 | PIXE-PIGME[12] | Good | Good | ? | Good | ? Good |

* = Some sources highly distinctive. - = Poor results.

[1]Green, 1962. [2]Reeves and Armitage, 1973. [3]Green et al., 1967. [4]Armitage et al. 1972. [5]Ward, 1972; 1974a; 1974b. [6]Coote et al., 1972; Bird et al., 1978. [7]Leach and Fankhauser, 1978. [8]Leach et al., 1978. [9]Leach, 1977a; 1977b; McCallum et al., 1979. [10]Leach, 1977a; 1977b. [11]Leach and Warren, n.d. [12]Duerden et al., 1979; 1980.

analysis. Recently, McCallum et al. (1979) have developed a technique which partly overcomes this difficulty. All element peaks are presented as ratios with respect to Fe, and the whole suite is cross-normalized against the Rb/Fe ratio (the concentrations of which are assumed to be linearly related). Unfortunately this has the undesirable effect of confusing the intra-source variations with the geometric effect. Consequently, some loss of precision has to be anticipated in identifying artefacts. In general, energy-dispersive XRF does not have as much discrimination as wavelength-dispersive XRF. In short, Ward's XRF results did not give sufficiently clear separation of sources for his element data to be simply applied to an energy-dispersive system. Further research in this field is underway, and in the future may be rewarding; however, it has not yet been proven that there is no overlap in the multivariate dispersion matrix between all pairs of known source populations.

It was with this general gackground that a programme of neutron activation analysis (NAA) was begun in 1976. This technique produces information on a large number of elements, and it was hoped that this knowledge could be used to devise a simpler technique or set of techniques which could be applied to obsidian artefacts.

## NEUTRON ACTIVATION ANALYSIS: BASIC RESULTS

An average of 10 samples from each of the known New Zealand sources, and a few Oceanic ones, were prepared for analysis. An attempt was made to set up a non-destructive analytical technique, by analysing duplicate powdered and whole samples; however, geometrical factors in the obsidian flakes caused some variation in results, and this procedure was abandoned in the interests of obtaining the most reliable quantitative results. The full details of this programme of work are described elsewhere (Leach and Warren, n.d.), and need only be summarized here.

About 34 isotopes were identified in the obsidian, but for various reasons only about 23 elements can be considered reasonably reliable. These run from $^{24}Na$ through to $^{238}U$ (determined from the isotope $^{239}Np$). The basic results of mean and standard deviation are given in Tables 11.2, 11.3, and 11.4. Close inspection of these figures shows that all sources surveyed may be reliably distinguished by their element compositions; however, one problem emerged which was not anticipated, and which casts some doubt on sourcing techniques which rely on only a few elements.

## THE PROBLEM OF CHATHAM ISLANDS OBSIDIAN ARTEFACTS

With notable exceptions, the element compositions of sources closest together geographically were found to be the most similar. This is not surprising considering the regional nature of New Zealand volcanic activity. The exceptions to this rule are Mayor Island obsidian, and three of the sources in the far north of New Zealand (Weta, Waiare, and Pungaere). These obsidians are clearly very similar to so-called 'Oceanic' obsidians associated with diverging or tensional plate boundaries. The remainder of the New Zealand obsidians show element compositions typical of 'circum-Oceanic' obsidians associated with converging plate boundaries (see Smith et al., 1977: 4, 10). The most notable element changes across this boundary in

TABLE 11.2 New Zealand and Oceanic obsidian sources studied, and isotopes identified. The units are those relevant to the columns in Tables 11.3 and 11.4

| SOURCES | | ISOTOPES | | UNITS |
|---|---|---|---|---|
| 1 | Weta | 1 | $^{24}Na$ | % |
| 2 | Waire | 2 | $^{46}Sc$ | ppb |
| 3 | Pungaere | 3 | $^{59}Fe$ | % |
| 4 | Huruiki | 4 | $^{60}Co$ | ppb |
| 5 | Burgess Island | 5 | $^{76}As$ | ppm |
| 6 | Awana | 6 | $^{82}Br$ | ppb |
| 7 | Te Ahumata | 7 | $^{86}Rb$ | ppm |
| 8 | Cooks Bay | 8 | $^{95}Zr$ | ppm |
| 9 | Purangi | 9 | $^{124}Sb$ | ppb |
| 10 | Hahei | 10 | $^{131}Ba$ | ppm |
| 11 | Tairua | 11 | $^{134}Cs$ | ppm |
| 12 | Maratoto | 12 | $^{140}La$ | ppm |
| 13 | Waihi red | 13 | $^{141}Ce$ | ppm |
| 14 | Waihi black | 14 | $^{147}Nd$ | ppm |
| 15 | Mayor Island green | 15 | $^{152}Eu$ | ppb |
| 16 | Mayor Island honey | 16 | $^{153}Sm$ | ppm |
| 17 | Mayor Island yellow | 17 | $^{160}Tb$ | ppb |
| 18 | Rotorua red | 18 | $^{175}Yb$ | ppm |
| 19 | Rotorua black | 19 | $^{177}Lu$ | ppb |
| 20 | Maraetai red | 20 | $^{181}Hf$ | ppm |
| 21 | Maraetai black | 21 | $^{182}Ta$ | ppm |
| 22 | Ongaroto | 22 | $^{233}Pa$ ($^{232}Th$) | ppm |
| 23 | Taupo | 23 | $^{239}Np$ ($^{238}U$) | ppm |
| 24 | Admiralty Islands | | | |
| 25 | Puu Waawaa (Hawaii) | | | |
| 26 | Maunga Orito (Easter Island) | | | |
| 27 | Motu Iti (Easter Island) | | | |
| 28 | Te Mamavai (Easter Island) | | | |
| 29 | Rano Kau (Easter Island) | | | |
| 30 | Pitcairn Island | | | |
| 31 | Otago glass | | | |
| 32 | Canterbury pitchstone | | | |
| 33 | Fanal Island | | | |

TABLE 11.3 Mean element concentrations of New Zealand and Oceanic obsidians from NAA. (NB: zero indicates unresolved peaks)

| | ISOTOPE | | | | | | | | | | | | | | | | | | | | | | |
|---|---|---|---|---|---|---|---|---|---|---|---|---|---|---|---|---|---|---|---|---|---|---|---|
| Sources | 1 | 2 | 3 | 4 | 5 | 6 | 7 | 8 | 9 | 10 | 11 | 12 | 13 | 14 | 15 | 16 | 17 | 18 | 19 | 20 | 21 | 22 | 23 |
| 1 | 3.3 | 2178 | 0.8 | 314 | 23 | 0 | 490 | 151 | 1286 | 239 | 27 | 16 | 34 | 29 | 0 | 9 | 1873 | 19 | 1159 | 6 | 4 | 48 | 12 |
| 2 | 4.5 | 267 | 3.2 | 317 | 56 | 0 | 661 | 1557 | 1906 | 498 | 46 | 151 | 225 | 132 | 423 | 24 | 5327 | 49 | 2734 | 48 | 24 | 100 | 28 |
| 3 | 4.4 | 243 | 3.2 | 469 | 13 | 703 | 656 | 1650 | 2075 | 491 | 39 | 145 | 213 | 123 | 348 | 26 | 5057 | 51 | 2801 | 48 | 25 | 96 | 21 |
| 4 | 3.4 | 4225 | 1.0 | 795 | 13 | 563 | 148 | 122 | 257 | 566 | 10 | 33 | 57 | 31 | 547 | 6 | 1029 | 8 | 530 | 5 | 1 | 12 | 3 |
| 5 | 1.3 | 7721 | 1.7 | 840 | 11 | 0 | 267 | 336 | 383 | 1900 | 12 | 47 | 80 | 38 | 1108 | 12 | 1289 | 12 | 812 | 12 | 1 | 14 | 3 |
| 6 | 2.7 | 4218 | 0.9 | 833 | 90 | 2815 | 190 | 158 | 204 | 675 | 9 | 37 | 62 | 34 | 538 | 6 | 1066 | 6 | 514 | 5 | 1 | 18 | 4 |
| 7 | 2.9 | 4285 | 0.9 | 750 | 6 | 845 | 212 | 122 | 0 | 524 | 11 | 39 | 69 | 34 | 401 | 7 | 1070 | 8 | 538 | 4 | 1 | 22 | 5 |
| 8 | 3.1 | 5435 | 1.0 | 588 | 5 | 878 | 123 | 106 | 0 | 731 | 4 | 29 | 49 | 25 | 785 | 5 | 778 | 6 | 423 | 4 | 1 | 13 | 3 |
| 9 | 3.7 | 5594 | 1.0 | 687 | 7 | 1674 | 134 | 122 | 147 | 813 | 4 | 32 | 51 | 26 | 850 | 5 | 782 | 7 | 485 | 4 | 1 | 13 | 3 |
| 10 | 3.1 | 4514 | 1.2 | 624 | 4 | 1794 | 142 | 161 | 231 | 936 | 6 | 33 | 58 | 28 | 857 | 6 | 873 | 8 | 501 | 4 | 1 | 14 | 2 |
| 11 | 3.0 | 5560 | 1.2 | 1354 | 11 | 2790 | 122 | 168 | 280 | 1032 | 6 | 24 | 40 | 15 | 818 | 4 | 725 | 5 | 380 | 5 | 1 | 11 | 3 |
| 12 | 3.2 | 3777 | 0.6 | 420 | 25 | 1084 | 185 | 83 | 395 | 947 | 10 | 27 | 44 | 20 | 303 | 4 | 633 | 5 | 370 | 3 | 1 | 16 | 4 |
| 13 | 3.0 | 4604 | 1.4 | 2536 | 4 | 1016 | 125 | 128 | 231 | 761 | 7 | 20 | 30 | 13 | 576 | 3 | 462 | 3 | 251 | 3 | 1 | 12 | 3 |
| 14 | 3.3 | 4564 | 1.4 | 2308 | 8 | 924 | 133 | 36 | 169 | 773 | 7 | 20 | 29 | 20 | 591 | 3 | 329 | 3 | 240 | 3 | 1 | 12 | 3 |
| 15 | 4.2 | 142 | 3.6 | 489 | 17 | 6322 | 143 | 960 | 247 | 54 | 5 | 95 | 163 | 103 | 1871 | 22 | 3576 | 21 | 1608 | 24 | 6 | 17 | 4 |
| 16 | 4.6 | 121 | 3.6 | 631 | 0 | 7915 | 139 | 1019 | 280 | 39 | 5 | 90 | 162 | 106 | 1688 | 20 | 3633 | 20 | 1588 | 24 | 6 | 16 | 4 |
| 17 | 4.1 | 106 | 3.2 | 523 | 0 | 0 | 139 | 975 | 200 | 53 | 4 | 91 | 166 | 107 | 1465 | 20 | 3382 | 19 | 1595 | 23 | 6 | 16 | 4 |
| 18 | 2.7 | 5576 | 0.9 | 804 | 27 | 2997 | 135 | 104 | 261 | 797 | 5 | 27 | 45 | 22 | 693 | 5 | 702 | 5 | 345 | 3 | 1 | 12 | 3 |
| 19 | 3.0 | 5660 | 0.9 | 1037 | 19 | 2933 | 131 | 146 | 235 | 851 | 5 | 28 | 47 | 26 | 763 | 5 | 666 | 5 | 376 | 4 | 1 | 12 | 2 |
| 20 | 2.9 | 3432 | 0.9 | 1260 | 0 | 5178 | 143 | 141 | 241 | 859 | 6 | 27 | 45 | 20 | 527 | 0 | 574 | 3 | 338 | 4 | 1 | 14 | 3 |
| 21 | 2.7 | 2991 | 0.8 | 1082 | 127 | 4531 | 140 | 141 | 212 | 859 | 6 | 24 | 40 | 17 | 492 | 3 | 506 | 3 | 308 | 3 | 1 | 13 | 3 |
| 22 | 3.1 | 3386 | 0.9 | 1144 | 40 | 1461 | 144 | 117 | 245 | 932 | 6 | 27 | 42 | 19 | 536 | 4 | 594 | 5 | 342 | 4 | 1 | 14 | 3 |
| 23 | 2.9 | 4251 | 0.9 | 1285 | 238 | 4178 | 124 | 166 | 205 | 683 | 5 | 24 | 38 | 22 | 602 | 4 | 571 | 4 | 323 | 4 | 1 | 11 | 3 |
| 24 | 3.3 | 3952 | 1.3 | 1439 | 0 | 3790 | 150 | 312 | 262 | 713 | 2 | 43 | 72 | 23 | 1030 | 0 | 1043 | 10 | 581 | 7 | 3 | 11 | 2 |
| 25 | 6.8 | 8790 | 3.3 | 107 | 0 | 1363 | 133 | 875 | 106 | 282 | 1 | 83 | 115 | 65 | 2821 | 11 | 1969 | 15 | 915 | 21 | 10 | 8 | 1 |
| 26 | 4.1 | 471 | 2.1 | 112 | 0 | 1866 | 70 | 800 | 240 | 495 | 1 | 96 | 179 | 101 | 3847 | 25 | 4667 | 32 | 1796 | 23 | 8 | 11 | 2 |
| 27 | 4.1 | 682 | 2.1 | 70 | 0 | 1927 | 61 | 634 | 231 | 459 | 1 | 89 | 155 | 93 | 3845 | 23 | 4339 | 29 | 1616 | 18 | 7 | 8 | 2 |
| 28 | 4.5 | 505 | 2.3 | 110 | 0 | 3042 | 80 | 797 | 256 | 509 | 1 | 105 | 181 | 110 | 4155 | 27 | 4991 | 35 | 1923 | 23 | 9 | 9 | 3 |
| 29 | 4.2 | 256 | 1.9 | 50 | 0 | 2008 | 79 | 743 | 257 | 466 | 1 | 106 | 185 | 109 | 3459 | 26 | 4740 | 34 | 1764 | 22 | 9 | 10 | 3 |
| 30 | 5.2 | 13601 | 4.8 | 1526 | 0 | 3378 | 89 | 729 | 344 | 2038 | 1 | 97 | 136 | 71 | 6003 | 18 | 2572 | 13 | 722 | 16 | 8 | 12 | 3 |
| 31 | 0.1 | 41952 | 0 | 323 | 0 | 0 | 66 | 0 | 0 | 287 | 8 | 28 | 55 | 40 | 1910 | 7 | 1698 | 18 | 0 | 2 | 0 | 5 | 8 |
| 32 | 3.1 | 7119 | 9.1 | 15199 | 0 | 541 | 99 | 714 | 177 | 789 | 1 | 137 | 155 | 120 | 6013 | 20 | 2484 | 14 | 0 | 14 | 11 | 15 | 0 |
| 33 | 2.9 | 3369 | 1.0 | 2174 | 1 | 0 | 205 | 118 | 596 | 378 | 14 | 29 | 49 | 20 | 364 | 0 | 878 | 8 | 475 | 4 | 1 | 20 | 1 |

TABLE 11.4 Standard deviations for element concentrations of New Zealand and Oceanic obsidians from NAA. (NB: zero indicates insufficient data).

| | ISOTOPE | | | | | | | | | | | | | | | | | | | | | | |
|---|---|---|---|---|---|---|---|---|---|---|---|---|---|---|---|---|---|---|---|---|---|---|---|
| Sources | 1 | 2 | 3 | 4 | 5 | 6 | 7 | 8 | 9 | 10 | 11 | 12 | 13 | 14 | 15 | 16 | 17 | 18 | 19 | 20 | 21 | 22 | 23 |
| 1 | 0.3 | 140 | 0.1 | 41 | 27 | 0 | 71 | 76 | 218 | 27 | 3 | 1 | 4 | 2 | 0 | 1 | 316 | 2 | 91 | 0 | 1 | 3 | 2 |
| 2 | 0.5 | 42 | 0.3 | 74 | 22 | 0 | 96 | 182 | 0 | 119 | 5 | 16 | 25 | 11 | 84 | 3 | 866 | 5 | 298 | 5 | 3 | 11 | 5 |
| 3 | 0.3 | 62 | 0.2 | 131 | 28 | 0 | 72 | 239 | 229 | 79 | 3 | 7 | 12 | 12 | 63 | 1 | 867 | 5 | 150 | 2 | 3 | 4 | 2 |
| 4 | 0.3 | 984 | 0.2 | 222 | 10 | 532 | 19 | 33 | 0 | 170 | 1 | 3 | 6 | 6 | 58 | 1 | 183 | 1 | 60 | 0 | 0 | 1 | 1 |
| 5 | 0.1 | 1512 | 0.3 | 71 | 9 | 0 | 85 | 5 | 0 | 322 | 3 | 6 | 6 | 0 | 88 | 1 | 32 | 1 | 0 | 1 | 0 | 2 | 1 |
| 6 | 0.3 | 1040 | 0.1 | 178 | 0 | 2383 | 35 | 38 | 83 | 178 | 2 | 3 | 7 | 8 | 309 | 1 | 315 | 1 | 94 | 1 | 0 | 2 | 2 |
| 7 | 0.5 | 728 | 0.1 | 175 | 0 | 332 | 46 | 26 | 0 | 85 | 1 | 6 | 8 | 6 | 69 | 1 | 161 | 1 | 79 | 1 | 0 | 3 | 1 |
| 8 | 0.4 | 700 | 0.1 | 73 | 0 | 213 | 21 | 24 | 0 | 125 | 0 | 4 | 6 | 3 | 99 | 1 | 116 | 1 | 50 | 0 | 0 | 2 | 1 |
| 9 | 0.2 | 283 | 0.1 | 59 | 0 | 825 | 17 | 38 | 29 | 75 | 0 | 1 | 2 | 4 | 64 | 0 | 122 | 0 | 18 | 0 | 0 | 0 | 0 |
| 10 | 0.2 | 275 | 0.1 | 54 | 7 | 734 | 9 | 31 | 41 | 190 | 0 | 2 | 2 | 3 | 92 | 1 | 179 | 1 | 32 | 0 | 0 | 1 | 0 |
| 11 | 0.4 | 728 | 0.2 | 310 | 1 | 2942 | 19 | 60 | 31 | 550 | 1 | 3 | 6 | 4 | 128 | 1 | 196 | 1 | 49 | 1 | 0 | 3 | 0 |
| 12 | 0.1 | 350 | 0.1 | 84 | 14 | 691 | 29 | 15 | 45 | 207 | 1 | 4 | 3 | 3 | 102 | 1 | 83 | 1 | 24 | 0 | 0 | 1 | 0 |
| 13 | 0.2 | 305 | 0.1 | 141 | 3 | 183 | 13 | 36 | 63 | 215 | 0 | 1 | 2 | 3 | 73 | 0 | 64 | 0 | 15 | 0 | 0 | 1 | 1 |
| 14 | 0.2 | 552 | 0.2 | 128 | 4 | 441 | 29 | 0 | 128 | 140 | 1 | 3 | 2 | 0 | 134 | 0 | 114 | 0 | 2 | 0 | 0 | 1 | 1 |
| 15 | 0.7 | 56 | 0.6 | 152 | 2 | 3367 | 33 | 166 | 40 | 34 | 1 | 15 | 21 | 16 | 394 | 4 | 544 | 5 | 227 | 4 | 1 | 3 | 1 |
| 16 | 0.8 | 32 | 0.6 | 107 | 0 | 2003 | 25 | 179 | 83 | 26 | 1 | 16 | 28 | 19 | 293 | 4 | 591 | 4 | 265 | 4 | 1 | 3 | 1 |
| 17 | 0 | 0 | 0 | 0 | 0 | 0 | 0 | 0 | 0 | 0 | 0 | 0 | 0 | 0 | 0 | 0 | 0 | 0 | 0 | 0 | 0 | 0 | 0 |
| 18 | 0.5 | 997 | 0.2 | 123 | 25 | 1950 | 31 | 40 | 25 | 320 | 1 | 4 | 6 | 3 | 152 | 1 | 112 | 2 | 45 | 0 | 0 | 2 | 1 |
| 19 | 0.5 | 1268 | 0.4 | 741 | 0 | 1934 | 28 | 57 | 67 | 276 | 1 | 5 | 7 | 6 | 223 | 1 | 213 | 1 | 73 | 1 | 0 | 2 | 1 |
| 20 | 0.4 | 484 | 0.2 | 383 | 0 | 2504 | 16 | 17 | 53 | 142 | 1 | 3 | 5 | 4 | 97 | 0 | 74 | 0 | 48 | 1 | 0 | 1 | 0 |
| 21 | 0.4 | 527 | 0.1 | 205 | 0 | 2960 | 29 | 23 | 27 | 221 | 1 | 4 | 6 | 4 | 86 | 0 | 79 | 1 | 59 | 1 | 0 | 2 | 0 |
| 22 | 0.5 | 400 | 0.1 | 172 | 55 | 516 | 14 | 31 | 43 | 322 | 1 | 3 | 2 | 2 | 39 | 0 | 50 | 0 | 10 | 0 | 0 | 1 | 0 |
| 23 | 0.4 | 710 | 0.2 | 179 | 321 | 2057 | 28 | 32 | 35 | 224 | 1 | 4 | 6 | 4 | 134 | 0 | 106 | 1 | 48 | 1 | 0 | 2 | 0 |
| 24 | 0 | 0 | 0 | 0 | 0 | 0 | 0 | 0 | 0 | 0 | 0 | 0 | 0 | 0 | 0 | 0 | 0 | 0 | 0 | 0 | 0 | 0 | 0 |
| 25 | 0 | 0 | 0 | 0 | 0 | 0 | 0 | 0 | 0 | 0 | 0 | 0 | 0 | 0 | 0 | 0 | 0 | 0 | 0 | 0 | 0 | 0 | 0 |
| 26 | 0.9 | 101 | 0.4 | 68 | 0 | 523 | 11 | 187 | 49 | 118 | 0 | 21 | 34 | 20 | 790 | 5 | 889 | 7 | 322 | 5 | 2 | 4 | 1 |
| 27 | 0.9 | 152 | 0.5 | 17 | 0 | 175 | 7 | 129 | 45 | 88 | 0 | 19 | 34 | 19 | 861 | 5 | 988 | 6 | 351 | 4 | 1 | 1 | 1 |
| 28 | 0.8 | 87 | 0.4 | 74 | 0 | 1705 | 18 | 100 | 39 | 106 | 0 | 19 | 31 | 19 | 728 | 6 | 1036 | 6 | 331 | 4 | 2 | 2 | 0 |
| 29 | 0 | 0 | 0 | 0 | 0 | 0 | 0 | 0 | 0 | 0 | 0 | 0 | 0 | 0 | 0 | 0 | 0 | 0 | 0 | 0 | 0 | 0 | 0 |
| 30 | 0 | 0 | 0 | 0 | 0 | 0 | 0 | 0 | 0 | 0 | 0 | 0 | 0 | 0 | 0 | 0 | 0 | 0 | 0 | 0 | 0 | 0 | 0 |
| 31 | 0 | 2427 | 0 | 253 | 0 | 0 | 0 | 0 | 0 | 18 | 0 | 2 | 0 | 8 | 69 | 0 | 249 | 0 | 0 | 0 | 0 | 5 | 0 |
| 32 | 0.8 | 902 | 1.1 | 2008 | 0 | 122 | 22 | 137 | 84 | 140 | 0 | 43 | 24 | 27 | 974 | 3 | 164 | 3 | 0 | 2 | 2 | 2 | 0 |
| 33 | 0.1 | 96 | 0.1 | 225 | 1 | 0 | 6 | 22 | 44 | 24 | 1 | 1 | 1 | 3 | 26 | 0 | 67 | 0 | 14 | 0 | 0 | 0 | 0 |

New Zealand are marked increases (in the case of 'Oceanic' obsidians) of no less than 13 elements ranging from Na to Ta (Na, Fe, Zr, La, Ce, Nd, Eu, Sm, Tb, Yb, Lu, Hf, and Ta). There is an equally marked decrease in three elements—Sc, Co and Ba.

We also analysed some archaeological specimens from the Chatham Islands (14 artefacts). This is a small island group about 600 miles (965 km) to the east of New Zealand. There is no known local source of obsidian, although there is an unsubstantiated early account of such a source below sea level. The prehistoric inhabitants of this island group are Eastern Polynesian in their physical characteristics, language, and material culture. It is not known with certainty though exactly where the founding population came from. The probable lack of a local source of obsidian raises the distinct possibility that the earliest inhabitants managed to bring some of the raw material with them, since the extreme isolation of this island group more or less rules out the possibility of two-way voyaging to another area in Polynesia. The identification of the natural source of these obsidian artefacts is therefore of some importance. Discovering the source of these archaeological artefacts has ultimately proven so difficult that we have been forced to re-think the whole identification procedure which is employed with this kind of work.

The Chatham Island artefacts are made from good quality obsidian which is green in transmitted light. There are two sources of green obsidian in nearby New Zealand—one on Mayor Island, and the other at Kaeo, in the far north. In 1973, Leach carried out some analyses of two Chatham Island artefacts using the Ward technique of wavelength dispersive XRF analysis and concluded (quite correctly) that the artefacts had a very similar trace element composition to the Mayor Island source, and were rather different to the Kaeo source. He mistakenly concluded on these grounds that Mayor Island was therefore the origin (Leach, 1973).

The more extended study by NAA has revealed that Mayor Island obsidian is extremely similar in element composition to that from Easter Island, some 4,300 miles (6,900 km) to the east. We now realize that it is simply not good enough to be able to show that an artefact is very similar to some nearby source, but that we should really be able to show that it is actually different to sources far away as well. The observed similarity between these distant sources may well be in keeping with current theory relating to plate tectonics in the Pacific; but for an archaeologist who wishes to document both long and short range prehistoric contact, as well as the origin of founder populations, this finding is a grim reminder to be very careful in determining the origin of prehistoric artefacts.

When the NAA results are looked at for 'differences' instead of 'similarities' an interesting picture emerges. By pooling the results together for the individual pieces from each of the three regions, and then comparing one group with another, a clear difference can be seen between any one type and another (see Table 11.5). It is noteworthy that all are green in transmitted light. Within the general perspective of Pacific obsidians these three are extremely similar, and even now it cannot be claimed with full confidence that the Chatham Island artefacts are from a third as yet undiscovered green source. When such closely similar results are obtained it pays to be more circumspect than ever

TABLE 11.5 A Comparison of Chatham Islands obsidian artefacts with source material from Mayor Island and Easter Island. Student's t test, 1% level (two tail test).

| ISOTOPE | Mayor Is Easter Is | Chatham Is Easter Is | Chatham Is Mayor Is. |
|---|---|---|---|
| $^{24}Na$ | - | - | - |
| $^{46}Sc$ | different | different | - |
| $^{59}Fe$ | different | different | - |
| $^{60}Co$ | different | - | different |
| $^{76}As$ | different | - | different |
| $^{82}Br$ | different | different | different |
| $^{86}Rb$ | different | different | - |
| $^{95}Zr$ | - | - | - |
| $^{124}Sb$ | - | - | - |
| $^{131}Ba$ | different | different | - |
| $^{134}Cs$ | different | different | different |
| $^{140}La$ | - | - | - |
| $^{141}Ce$ | - | - | - |
| $^{147}Nd$ | - | - | - |
| $^{152}Eu$ | different | different | - |
| $^{153}Sm$ | - | - | - |
| $^{160}Tb$ | different | - | - |
| $^{175}Yb$ | different | - | different |
| $^{177}Lu$ | - | - | - |
| $^{181}Hf$ | - | - | - |
| $^{182}Ta$ | - | - | - |
| $^{233}Th$ | different | different | - |
| $^{238}U$ | - | - | - |

Sample sizes: Mayor Island N=10, Easter Island N=6, Chathams Islands N=14.

TABLE 11.6 Obsidian Artefact Source Identification Schema

STEP 1: CHARACTERIZATION OF ALL SOURCES ACROSS THE FULL RANGE OF VARIATION AS DISTINCTIVE FROM EACH OTHER.

STEP 2: DETERMINE THE SAME PROPERTIES ON THE ARTEFACT.

STEP 3: DECIDE WHICH SOURCE THE ARTEFACT IS DERIVED FROM.

METHOD 1: (William of Occam) Which is the most similar source? Minimum Euclidean distance technique, such as finding the smallest discriminant function. This method will fail to give correct answers when there is overlap in the multivariate dispersion matrixes of sources, or when there are undiscovered sources. This method is not advised.

METHOD 2: (Karl Popper) Which sources is the artefact different to?

Case 1: Only one possible source remains—carry out Hottelings T test. No significant difference implies that the probable but not certain source has been found.

Case 2: More than one possible source remains—refine Step 1.

about the possibility of sample bias from sources, problems of experimental replication, and so on. Further research into some of the lighter elements is proceeding, and it is hoped to finally resolve this matter in the near future.

## A COMMENT ON THE GENERAL PROCESS OF ARTEFACT SOURCING

The example of the Chatham Island artefacts is instructive from several points of view, and has made us re-think the whole process of source identification.

There are three distinct steps involved in identifying the original source of prehistoric artefacts (see Table 11.6). Firstly, by some technique or another we must be able to show that all potential sources involved can be effectively distinguished across the full range of variation within each source. In this respect, it has to be admitted that few techniques explored so far in New Zealand have got past first base. The Chathams example suggests that it would be wise to extend this characterization phase beyond the immediate confines of the prehistoric people in any one island group. Secondly, the same properties should be assessed for the artefacts under identical experimental conditions.

It is in the third step that some difficulties might be encountered. The usual method for deciding which source the artefact originally came from is by employing Occam's razor, that is, by finding the most parsimonious solution. This is most commonly achieved by using statistical techniques such as discriminant function analysis, or by one of a variety of minimum Euclidean distance functions. In our opinion, serious errors can be introduced by slavishly following these 'similarity' functions. For one thing, assignments of low probability may be due either to true intra-source variability, or to the presence of unstudied obsidian sources. These two alternatives are very difficult to distinguish. As Leach's study has shown (1973), the Chatham Islands artefacts were clearly ascribed to the Mayor Island source with a respectably high probability when this procedure was followed; yet this conclusion may now be questioned.

In our opinion it is wiser to adopt a rather more critical approach of trying to prove where the artefact did not come from, and systematically go through all the sources, including those from further afield. This more rigorous approach is closer to that advocated by Karl Popper, where a genuine attempt is made at refutation rather than seeking the simplest solution (Occam's razor). By concentrating on differences rather than similarities, we should be more certain of the source identifications which are made. In addition, we should be in a better position to recognize material from new sources when it turns up archaeologically.

The original purpose of carrying out the NAA was to try and devise a simple technique or set of techniques for sourcing New Zealand obsidian artefacts—what then can be suggested?

## A STRATIFIED SOURCING TECHNIQUE FOR NEW ZEALAND OBSIDIANS

In devising an identification scheme, the original criteria were firmly kept in mind: that is it should be non-destructive, reliable, rapid and not too complex, and readily available to archaeologists for running thousands

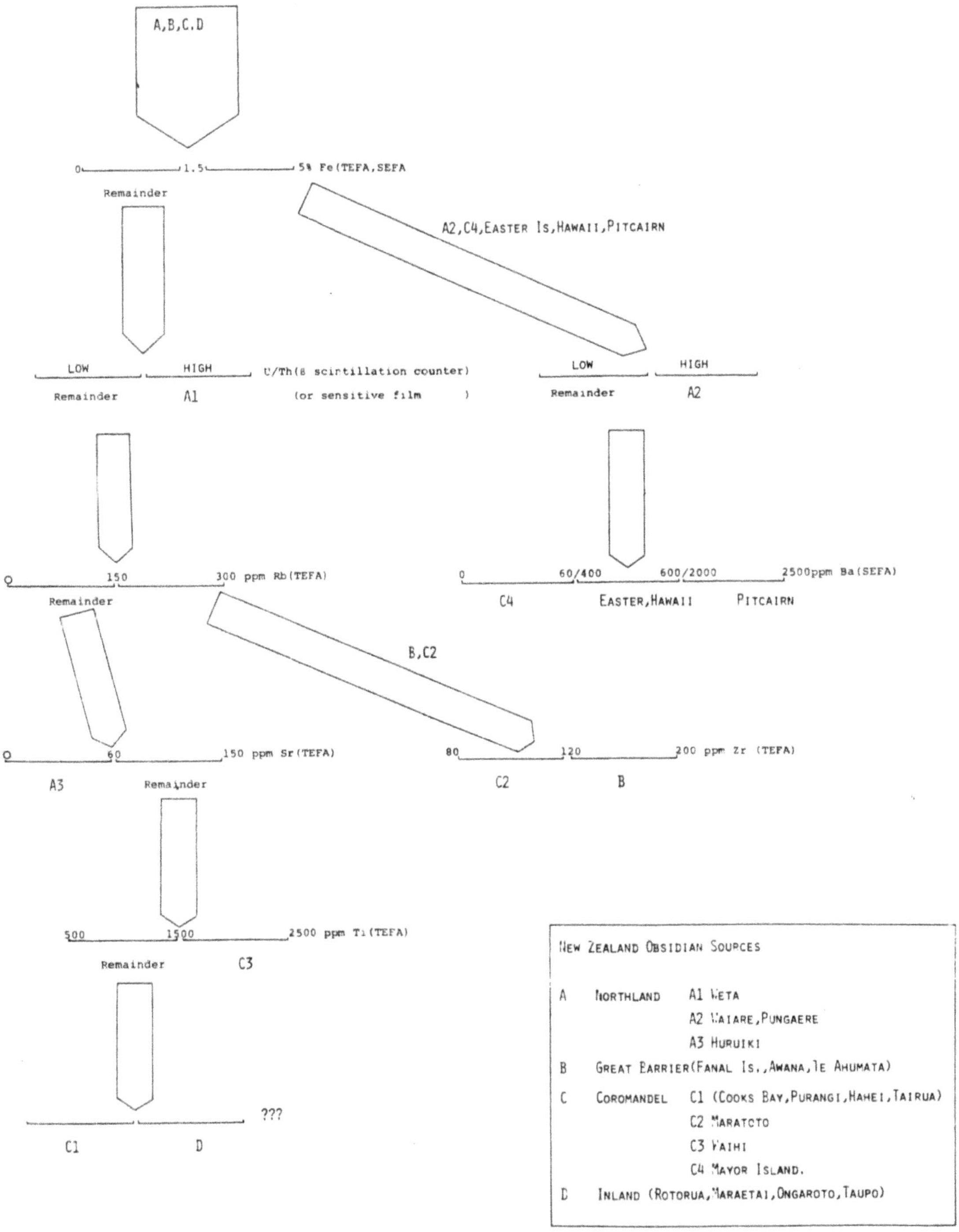


**Figure 11.2 Stratified sourcing schema for New Zealand obsidians.**

of samples at minimal cost. It should be noted that although desirable, it is not considered necessary to distinguish between all of the 18 New Zealand sources. They are geographically grouped into four regions (see Figure 11.1), and most archaeologists would be perfectly happy to know which regions of supply are involved in an archaeological site, so long as they could be certain of the results.

The minor and trace element data have been carefully examined to see which sources may be individually screened out as 'different' from the remainder with near 100% certainty. In some cases, a particular element concentration for one source is different to all other sources by an order-of-magnitude, and consequently may be screened out with considerable reliability. The schema suggested (see Figure 11.2) relies on the use of two instruments.

1) Energy-dispersive XRF which can operate on both modes of excitation —that is by primary X-ray tube, and also by radioactive source. The tube is used for the analysis of Fe, Rb, Sr, Zr and Ti, and the radioactive source is needed for the identification of the K alpha peak of Ba. This element requires somewhat greater excitation energy than can be provided by a normal X-ray tube, Source excitation analysis normally takes quite some time, but the differences in concentrations of Ba are such that qualitative assessment can be made within a few minutes.

2) A method for distinguishing certain obsidians which have far more natural radioactivity than all others. There are two simple ways of achieving this separation. There is a significant difference in natural beta particle emission, and this can be taken advantage of using a liquid scintillation counter (see Leach et al., 1978). An even simpler method would be to use an alpha sensitive film (Kodac LR-115, 13 mμ emulsion). Large batches of obsidian artefacts are simply left on small pieces of the film and later developed. Although this is not yet a proven technique, it is expected that the large difference in natural radioactivity will enable simple screening on the basis of high and low density of tracks recorded on such film.

The identification schema has one major problem. As yet, no simple technique can be suggested which can distinguish the inland group of sources (D) from those of the north-east Coromandel region (C1) Until such a simple method is developed, the few obsidian artefacts which end up in this mixed group will have to be identified by one of the more complex methods such as PIXE analysis (Duerden et al., 1979: 1980).

CONCLUSIONS

Obsidian artefacts are ubiquitous in New Zealand archaeological sites of all periods, and offer a highly suitable means of determining prehistoric communication patterns. No less than 12 different attempts have been made to devise a sourcing method so that the natural origin of prehistoric artefacts can be identified. Four of these have been shown to be successful, but so far only about 500 actual artefacts have been identified. All techniques examined, including these four, have serious drawbacks as far as routine application by any archaeologist is concerned. They are either too expensive for large runs of artefacts, or too slow, or require partial or complete destruction of artefacts. The information gained from these various attempts, together with that from a large programme of NAA has been used to suggest a much simpler scheme of analysis. The fact that a perfect scheme still eludes us is hardly surprising—it is after all a complex problem in New Zealand. However, if such a scheme were followed, it is believed that only a small portion of artefacts would end up in the final screening tray; and in the meantime these will have to be identified by a more complex technique. With further research, this final group will doubtless yield to a simple separation technique also.

REFERENCES


Armitage, G. C., Reeves, R. D. and Bellwood, P. S., 1972. Source identifications of archaeological obsidians in New Zealand. New Zealand Journal of Science 15: 408-420.

Bird, J. R., Russell, L. H., Scott, M. D. and Ambrose, W. R., 1978. Obsidian characterization with elemental analysis by proton induced gamma-ray emission. Analytical Chemistry 50(14): 2082-2084.

Coote, G. E., Whitehead, N. E. and McCallum, G. J., 1972. A rapid method of obsidian characterization by inelastic scattering of protons. Journal of Radioanalytical Chemistry 12: 491-496.

Duerden, P., Cohen, D. D., Clayton, E., Bird, J. R., Ambrose, W. R. and Leach, B. F., 1979. Elemental analysis of thick obsidian samples by proton induced X-ray emission spectrometry. Analytical Chemistry 51(14): 2350-2354.

Duerden, P., Bird, J. R., Scott, M. D., Clayton, E., Russell, L. H. and Cohen, D. D., 1980. PIXE-PIGME studies of artefacts. Nuclear Instruments and Methods 168: 447-452.

Green, R. C., 1962. Obsidian: its application to archaeology. New Zealand Archaeological Association Newsletter 5: 8-16.

Green, R. C., Brooks, R. and Reeves, R. D., 1967. Characterization of New Zealand obsidians by emission spectroscopy. New Zealand Journal of Science 10: 673-682.

Leach, B. F., 1973. Obsidian from the Chatham Islands. New Zealand Archaeological Association Newsletter 16(3): 104-106.

Leach, B. F., 1977a. Progress towards the routine sourcing of New Zealand Obsidians. New Zealand Archaeological Association Newsletter 20(1): 6-17.

Leach, B. F., 1977b. A rapid method of sourcing New Zealand lithic material using a low power XRF analyser. Oceanic Prehistory Records 3.

Leach, B. F. and Fankhauser, B., 1978. The characterization of New Zealand obsidian sources by use of thermoluminscence. Journal of the Royal Society of New Zealand 8(4): 331-342.

Leach, B. F., Warren, S. E. and Fankhauser, B., 1978. Obsidian from the far north of New Zealand: a method of sourcing based on natural radioactive emission. New Zealand Journal of Science 21: 123-128.

Leach, B. F. and Warren, S. E., n.d. The chemical composition of New Zealand and Oceanic obsidians from neutron activation analysis. Manuscript, Otago University.

McCallum, G. J., McFadgen, B. G., Moore, P. R., Richardson, L. M. and Shèppard, R. A., 1979. Characterization of New Zealand obsidian by energy dispersive X-ray fluorescence. Manuscript 1001, Institute of Nuclear Sciences, New Zealand.

Reeves, R. D. and Armitage, G. C., 1973. Density measurements and chemical analysis in the identification of New Zealand obsidians. New Zealand Journal of Science 16: 561-572.

Shutler, R. and M. E. Shutler, 1975. Oceanic Prehistory. Cummings.

Smith, I. E. M., Ward, G. K. and Ambrose, W. R., 1977. Geographic distribution and the characterization of volcanic glasses in Oceania. Archaeology and Physical Anthropology in Oceania 12(3): 173-201.

Ward, G. K., 1972. Obsidian and New Zealand archaeology: a paradigm for sourcing artefact obsidians using X-ray fluorescence spectrography. MA Thesis, Anthropology, University of Otago.

Ward, G. K., 1974a. A paradigm for sourcing New Zealand archaeological obsidians. Journal of the Royal Society of New Zealand 4: 47-62.

Ward, G. K., 1974b. A systematic approach to the definition of sources of raw material. Archaeometry 16: 41-53.

# 12. TECHNOLOGICAL CHANGES IN THE DEVELOPMENT OF POLYNESIAN ADZES

Helen M. Leach

Before archaeology came of age in Oceania, adzes were seen as culture markers from which past human migrations could be traced. Thus a combination of taxonomic and distributional analyses occupied early scholars who worked with museum collections of uncertain age. They saw a very obvious contrast between lenticular-sectioned, oval and planilateral axe-adzes found from Fiji to New Guinea, and the triangular and quadrangular adzes of Polynesia which often displayed a tang or butt modification.

Duff (1959) regarded the Polynesian types as directly influenced from the Philippines area in the case of the thick quadrangular, as a late Polynesian invention in the case of the triangular adze (apex to back) and as a "Polynesian elaboration on an ancestral theme" when considering the triangular adze (apex to front). In his later work (1970), he argued for even more Southeast Asian influences. Essentially these are historical explanations of the contrast between Melanesian and Polynesian adzes, based on the age-area theory of distribution.

Although primarily interested in taxonomy, Skinner in 1938 had speculated on how far materials might force the adze maker to modify his types. He found the influence of material at the varietal but not the type level (Skinner, 1974). Green and Dessaint (1977: 17-18) have explored this field more rigorously and concluded:

> "that greenstone, greywacke, and shell do affect one or more of the size measurements on adzes made in these materials. But these differences are not necessarily reflected in morphological restrictions on the forms of adzes made in these materials. Rather one can present strong arguments for the importance of the craftsman's intentions and various cultural preferences to explain the range of types produced".

Elsewhere, in examining innovation in the Polynesian adze kit after the settlement of Samoa, Green argues that:

> "a restriction in the durable rock types and very large clam shells available for adze manufacture was imposed by the change in location. This, I believe, would have had an effect on the adze types produced" (Green, 1974b: 264).

If Green is uncertain how materials affect shape, as these quotations suggest, it is probably because Pacific prehistorians have studied finished form and raw material but not the technology that transforms one into the other.

Functional explanations have been offered for chronological changes in the New Zealand adze kit. Several workers have described Archaic adzes as specialist canoe-making tools, which lost their importance as canoe styles changed and forest clearance for horticulture and palisade building gained momentum (Best, 1977). The same argument might be applied in West Polynesia: that a desire for larger ocean-going canoes and tools to make them, arose in Samoa and culminated in the settlement of East Polynesia. This view would underestimate the obvious voyaging skills of Lapita groups in the western Pacific.

Best (1977: Figure 1) also offered a functional explanation of the stepped or offset butt, which is to protect the lashings from striking the worked surface during the follow-through. This is only a danger in thick quadrangular and high triangular (apex to front) adzes, which Best found to operate with a very low attack angle. His experiments showed that low attack angle in the quadrangular adze correlated with the fine shaving of flat wood surfaces. The same angle in a triangular adze with its narrow cutting edge might indicate the fine shaving of curved surfaces.

Obviously many types of explanation could be offered in addition to functional, historical or geological ones. To sort them out we require more information on the flaking properties of the various rock types, the form in which the raw material occurs (cobbles, boulders, beds with cleavage planes etc.), the functional value of the rock types (toughness, elasticity etc.), the performance of the new adze types in various tasks, and the technological steps involved in the manufacture of the new types.

Green (1971) has written a valuable paper on the chronology of these developments. Tangs were experimented with by Fijian Lapita groups, again in Samoa (A.D. 0-300) and more seriously in the Hane site in the Marquesas (A.D. 300-1000), where the first true tang appears. The thick quadrangular adze form also appears at Hane. The triangular adze (apex to front) occurs untanged in Samoa (A.D. 0-300) and first shows a modified butt at Maupiti (A.D. 800-1100). The triangular adze (apex down) may have a Samoan prototype (Green, 1969: 131-2) but is not a distinct type until the Maupiti 'period'. We seem to be dealing with a series of modifications and innovations spread over many centuries, so that the search for a single explanation is probably quite unrealistic. In addition, it should not be forgotten that different East Polynesian island groups concentrated on the production of different adze shapes often to the exclusion of others. This fact should be remembered by those hoping to correlate a specific activity with a particular type of adze. The Hawaiians obviously managed well with an adze kit dominated by quadrangular forms (Emory, 1968), while the Tahitians carried on presumably similar tasks with triangular varieties.

The manufacturing process is the greatest unknown, and it occupies a vital position between the raw material and the functional tool. Both material and function impinge on adze-making technology, in that some procedures may not be successful when carried out in basalt or greywacke, for example, and that some adze shapes easily produced from a technological point of view may not have the necessary qualities of weight or attack angle to accomplish a certain task.

The adze-making process begins at the quarry and often all the definitive steps in shaping the adze take place in the vicinity, on working floors. Prehistoric quarries and working floors are known at several localities in New Zealand (See Figure 12.1). Important ones are the Tahanga Hill basalt quarry on the Coromandel (Shaw, 1963; Moore, 1976; Best, 1977), the D'Urville Island and Nelson metasomatised argillite quarries (Skinner, 1913; Thomson, 1918; Duff, 1946; Keyes, 1961; 1975; Walls, 1974), the Brighton Island basalt workshop in Otago (von Haast, 1880), and the Riverton and Bluff Harbour argillite quarries in Southland (Leach and Leach, 1980; Huffadine, 1978). Material from each of these has been deposited in museums, and excavations have been conducted on working floors at Tahanga, Riverton, and Tiwai Point (Bluff Harbour).

In Hawaii, the Mauna Kea quarry and workshops are under investigation, the most systematic study yet devised for this type of site (McCoy 1977; McCoy and Gould 1977). Metraux (1940:273) noted the presence of an adze quarry on Easter Island. Verin (1969:175) described various workshops on Rurutu, Austral Is. On Samoa two quarries are recorded but may now have been destroyed (Green, 1974a:141). Although many have been destroyed, others probably remain undiscovered beneath dense vegetation in more remote areas of Polynesia. Before they vanish it is essential that a thorough examination is made of the form in which the raw material occurs as it may have an important influence on technological procedures.

In the Archaic site at Riverton, waterworn boulders of argillite were obtained from the foot of a sea cliff and split along internal fracture planes. Some slabs were levered off the weathered outcrop face but generally material without a thick cortex was selected. Small waterworn cobbles were also picked up along the beach. Lenticular adzes were manufactured from these cobbles (Figure 12.2) by flaking from both edges on to both faces (i.e., bidirectional bilateral flaking). The presence of large blocks with two parallel fracture plane surfaces facilitated the manufacture of large quadrangular adzes by guiding the trimming flakes into a right-angle mode. Blocks with a single fracture plane supplied a suitable striking platform for large thick flakes which could be made into either sub-triangular or trapezoidal-sectioned adzes, depending on the surface geometry of the front of the flake (Figure 12.3). Where sharp angles occurred between fracture planes and flake surfaces, the striking of triangular blades was facilitated, from which small triangular adzes or chisels were made (Figure 12.4). At this site, then, the surface geometry of the primary blocks had a profound effect on the size and shape of the adzes.

It is interesting to note that on the Mauna Kea site, the Hawaiians:

> "extracted the best available basalt by using levers and breaking up large slabs of rock that had already been cracked by thermal expansion and contraction" (McCoy and Gould, 1977: 238).

The slabs were quarried from depths up to 15 feet to obtain unweathered material, and their parallel sides were probably essential to the production of the quadrangular adze. It is possible that the dominance of this cross-section in Hawaii may be attributed to the form in which the raw material occurs. On Rurutu, Vérin noted that in certain cases the stone workers made use of fragments of dykes already split off by natural processes (Vérin, 1969: 177).

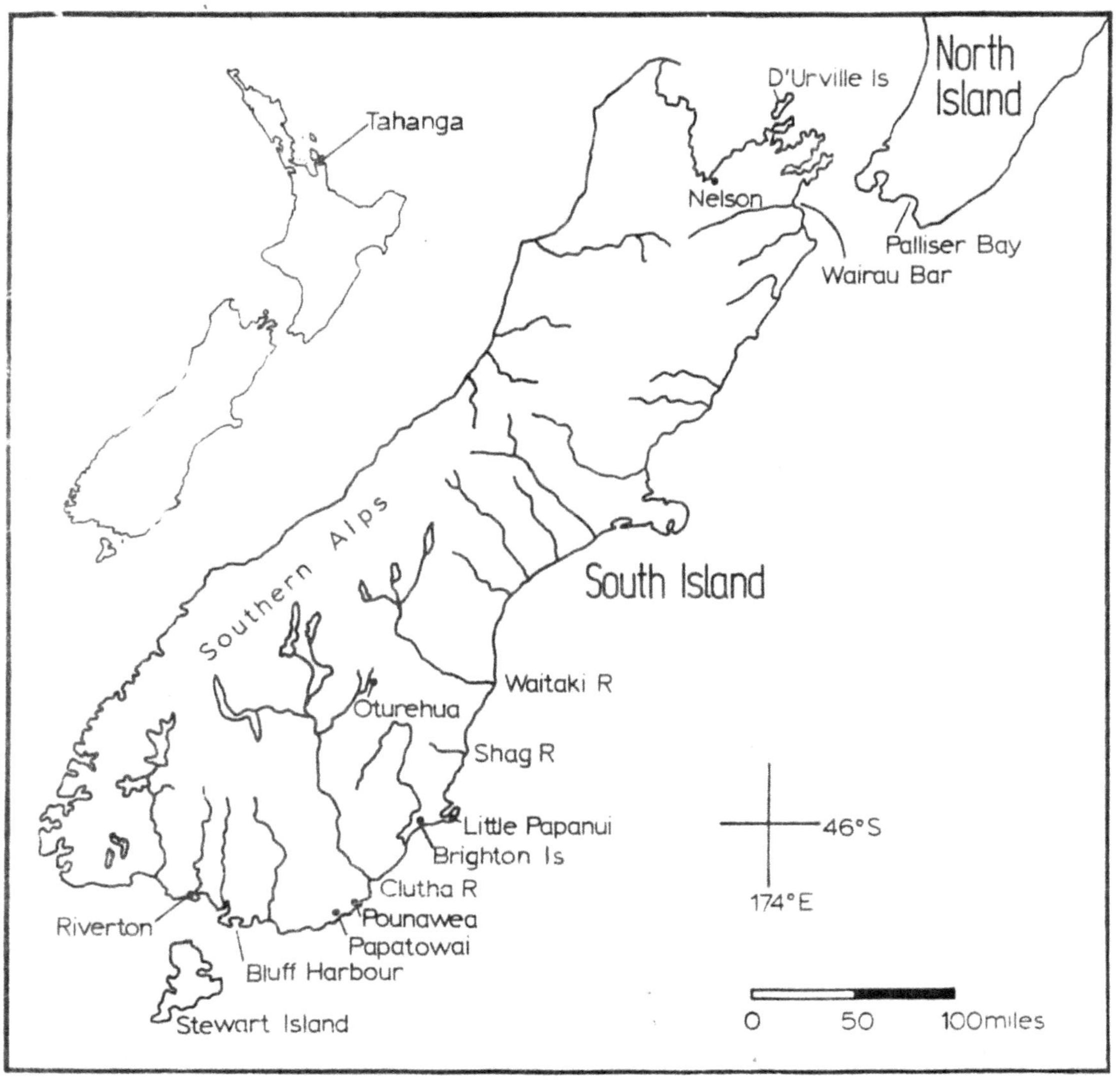


Figure 12.1 Location map: South Island, New Zealand, showing quarries and archaeological sites.

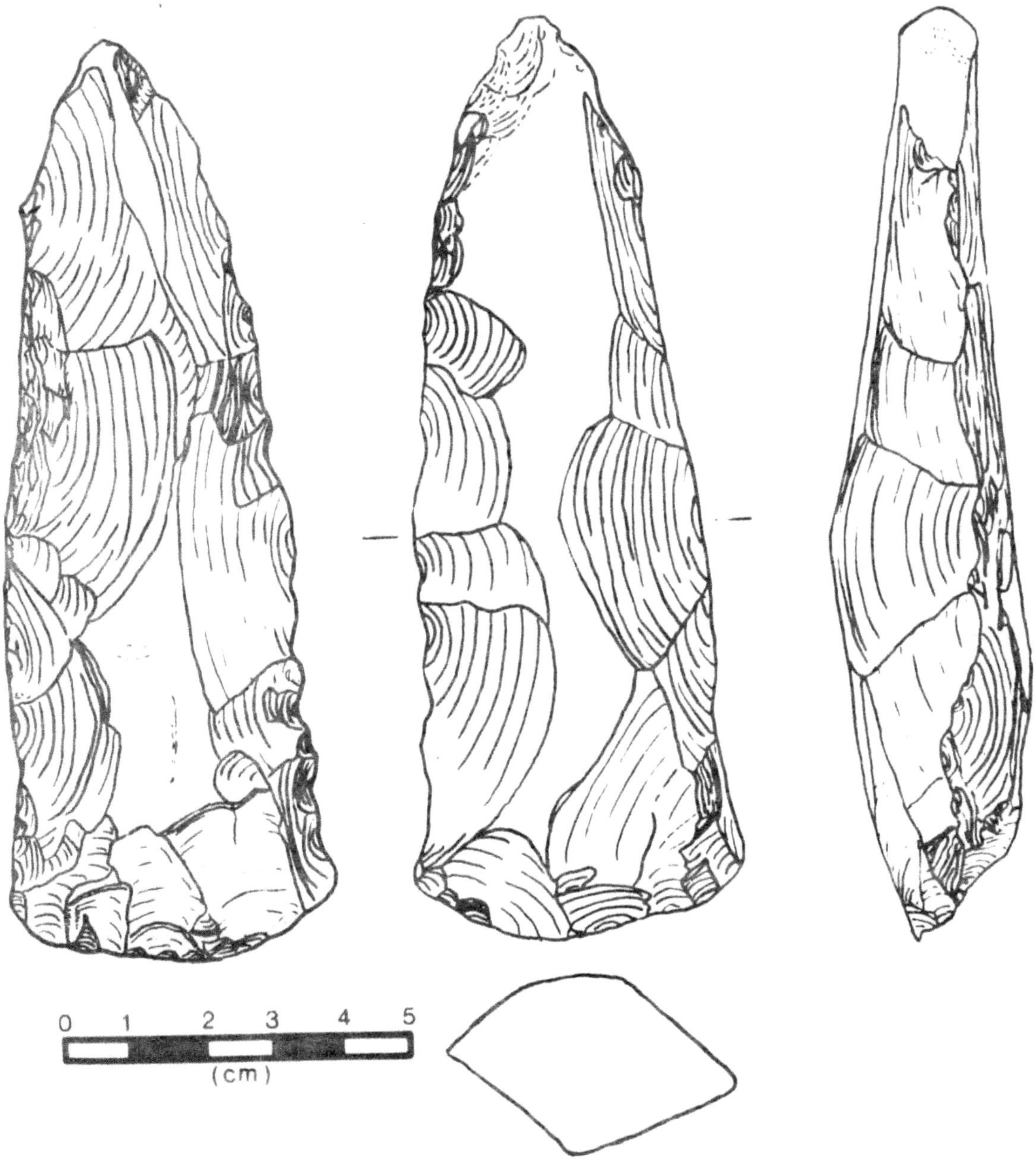


Figure 12.2 Lenticular preform made from beach cobble by bidirectional bilateral flaking. Riverton Z15 L-3A/B

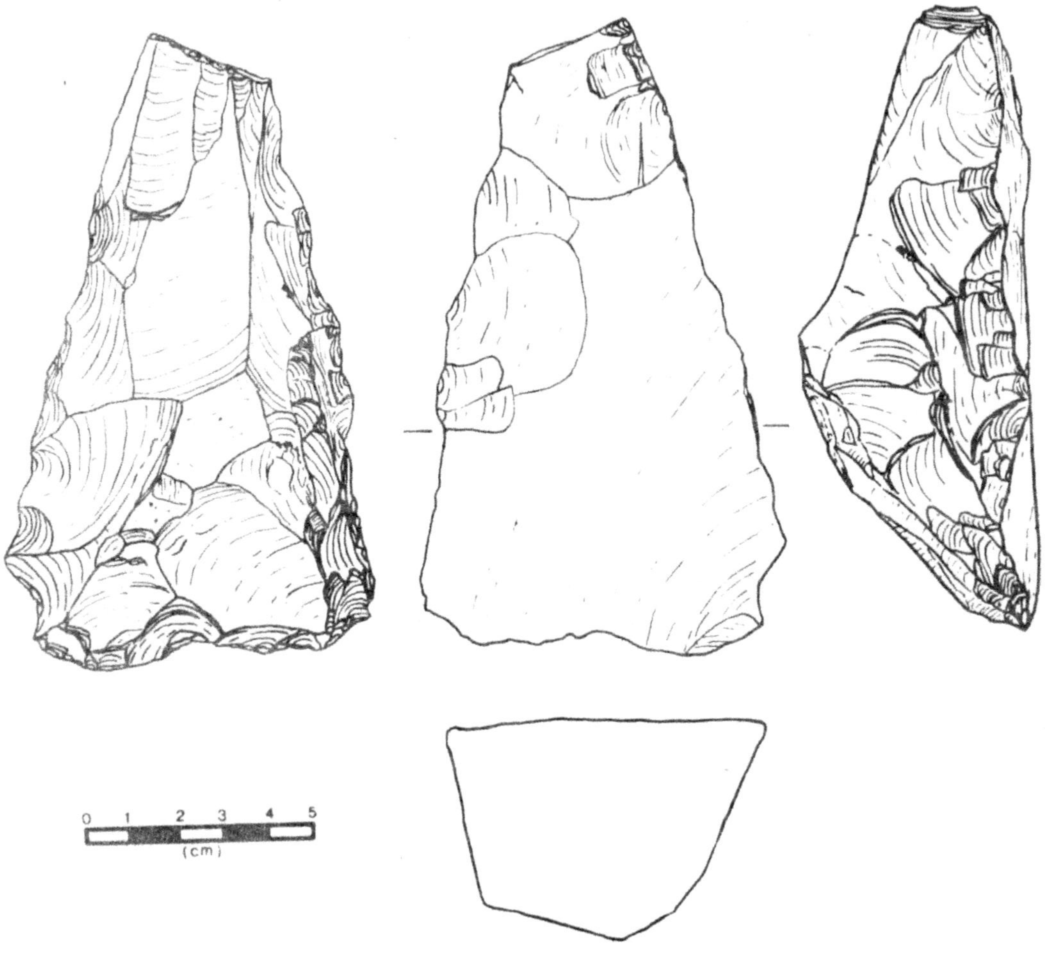


Figure 12.3 Sub-triangular preform made from thick flake. Riverton C2 L-3

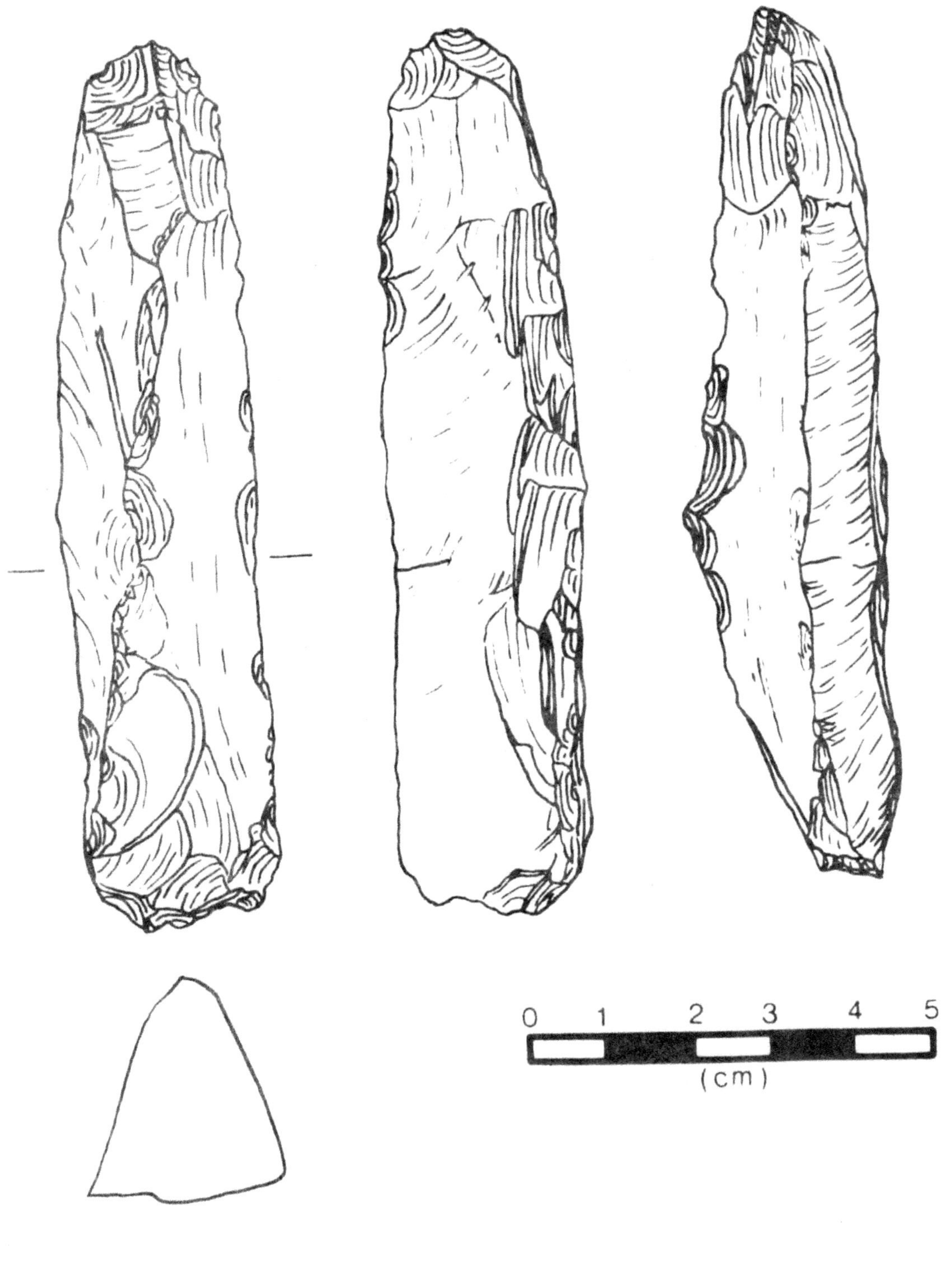


Figure 12.4 Triangular preform made from blade. Riverton AB 27 (B2)

Analysis of the Riverton workshop assemblage has revealed much new information about the technology of adze manufacture, in particular the shaping of the preform. Existing morphological classifications are based on finished cross-section, oriented with reference to the bevel. Many preforms were discarded before the bevel was made and surface ridges reduced, and thus cannot be classified. A new classification was devised (Figure 12.5) based on the number of edges from which trimming flakes were struck (in the case of a flake preform those flakes removed after the preform is detached from the parent block), and whether the flaking was uni- or bidirectional.

It was found that the most skilfully-made adzes were quadrilateral, made by striking high-angle flakes from four edges either unidirectionally or bi-directionally. Trilateral adzes were made by striking high angle flakes from three edges. Bilateral adzes, the best represented among the rejects, involved low-angle percussion from two edges. 'Jig-saw' analysis by core reconstruction showed that bilateral adzes were often made from the larger flakes detached during the manufacture of a quadrangular adze.

One of the most significant facts revealed by the application of this technological classification was that the triangular cross-sectioned adze (apex to front) can be produced by both trilateral flaking of a core, flake or blade or bilateral flaking of a thick flake (Figures 12.6 and 12.7). Thick quadrangular adzes are usually made by quadrilateral flaking of a slab or massive flake. Thin quadrangular or trapezoidal adzes are more commonly made by bilateral working of a flake. The various methods of achieving a similar final product may have considerable diagnostic value in separating out the threads of East Polynesian settlement and influences, for they offer the chance of distinguishing in the preform stage what is undistinguishable in the fully ground condition.

Because high-angle percussion is more difficult, core preparation prior to the removal of the preform is often evident, usually involving platform and ridge preparation techniques similar to those necessary for prismatic blade manufacture. It was found that the apex of several triangular preforms had been prepared by alternate flaking on either side of the ridge before the preform was struck from the parent block. This technique is evident in the 11th century blade manufacturing site of Oturehua in Central Otago (Leach, 1969).

The reduction of platform overhang which was practised by the successful blade makers of Oturehua is also found on both preforms and trimming flakes at Riverton, especially those with high angles between striking platform and the flake's dorsal surface. It can be recognized by multiple tiny step fractures just below the platform edge (Figure 12.8). Edge reduction hammers which performed this task have been identified from both sites, and display a characteristic battered notch. Percussion hammers are also present, usually oval water worn stones of various weights (Figure 12.9).

The relationship between blade and adze making is just one of the technological issues which can profitably be studied in Polynesia and Melanesia. Basalt blades have been found as waste flakes on Mauna Kea, Hawaii (McCoy and Gould, 1977: 242). Obsidian blades and micro-cores of polyhedral type are described from Easter Island (McCoy, 1976: 330-335). Gerard (1976) identified several blade tools from Eiao, Marquesas, including four blades with the bulbar surface untouched and two with it extensively trimmed. One

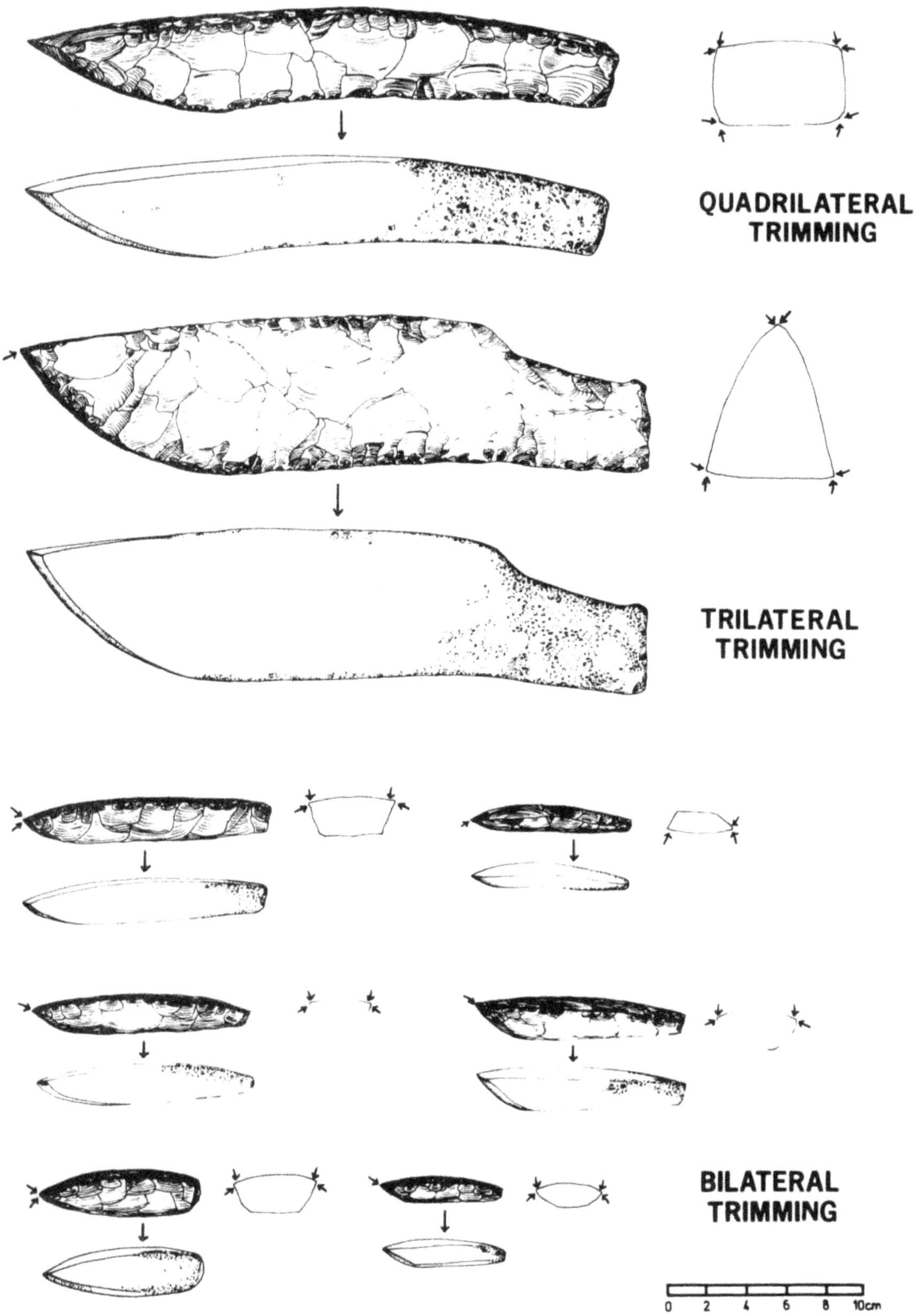


Figure 12.5 Classification scheme of adze preforms based on the number of edges from which trimming flakes are struck: quadrilateral, trilateral and bilateral trimming.

# TRIANGULAR ADZE (Apex up)

POSSIBLE PRODUCTION SEQUENCES

A. FROM CORE

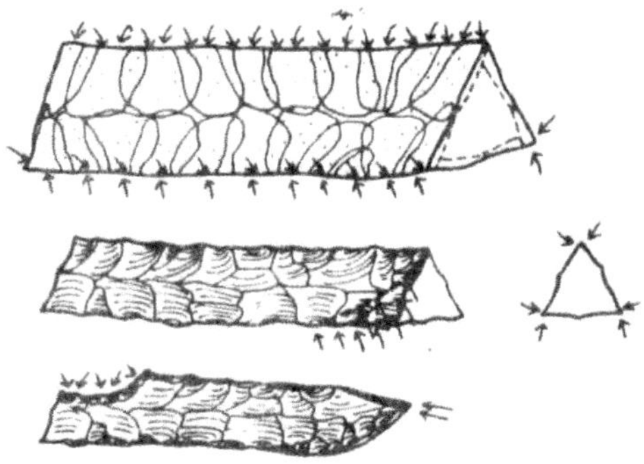

1. SELECTION OF RAW MATERIAL (E.G. FROM PRISMATIC DYKE)
2. BIDIRECTIONAL FLAKING FROM 3 CORNERS
3. SHAPING BEVEL BY REDUCING BACK FROM SIDES
4. REDUCTION OF APEX RIDGE TO MAKE TANG
5. REMOVAL OF 'BEAK' AND FORMATION OF NARROW CUTTING EDGE BY STRIKING OFF MICROBLADES

B. FROM FLAKE

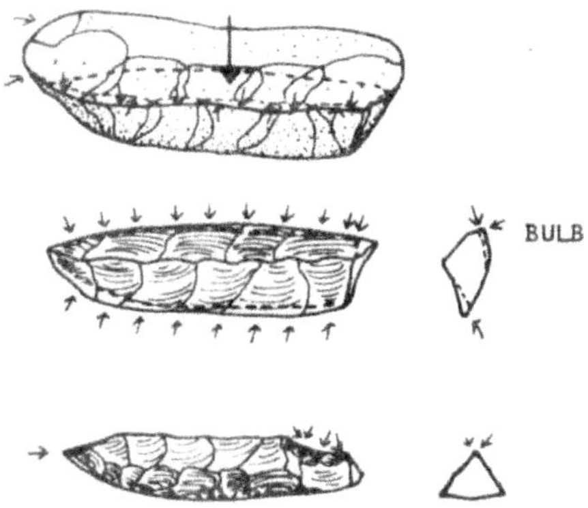


1. REMOVAL OF CORTEX FROM PARENT BLOCK, WORKING FROM AN EXISTING RIDGE
2. REMOVAL OF MASSIVE FLAKE WITH LONGITUDINALLY AND TRANSVERSELY CONVEX BULBAR SURFACE
3. STRAIGHTENING OF BACK AND SIDES AND SHAPING OF BEVEL WITH BILATERAL TRIMMING FLAKES STRUCK FROM SIDES
4. REDUCTION OF RIDGE FOR TANG AS IN A
5. SHAPING OF CUTTING EDGE AS IN A

C. FROM BLADE

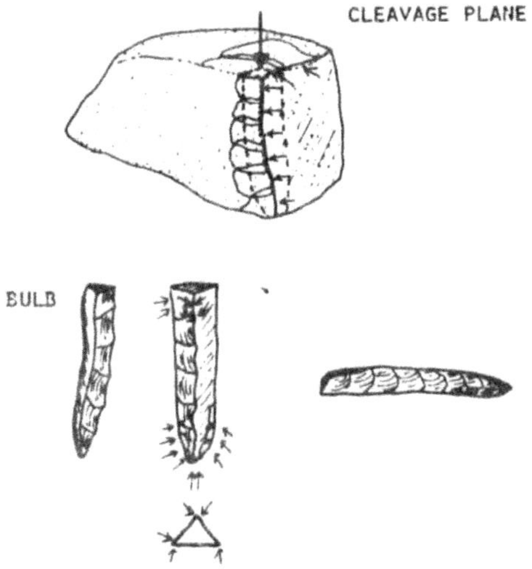


1. SELECTION OF RAW MATERIAL WITH EXISTING RIDGE
2. REMOVAL OF CORTEX FROM PLATFORM AREA AND FROM SIDE OF RIDGE
3. REMOVAL OF TRIANGULAR BLADE WITH LONGITUDINALLY CONCAVE BACK
4. REMOVAL OF BULB OF PERCUSSION WITH TRIMMING FLAKES STRUCK FROM SIDES
5. SHAPING OF BEVEL AND CUTTING EDGE AND TANG AS IN A

Figure 12.6 Possible production sequences for triangular adzes (apex up). N.B. In A, trilateral trimming occurs. B and C utilize a prepared core technique followed by bilateral (sometimes trilateral) trimming.

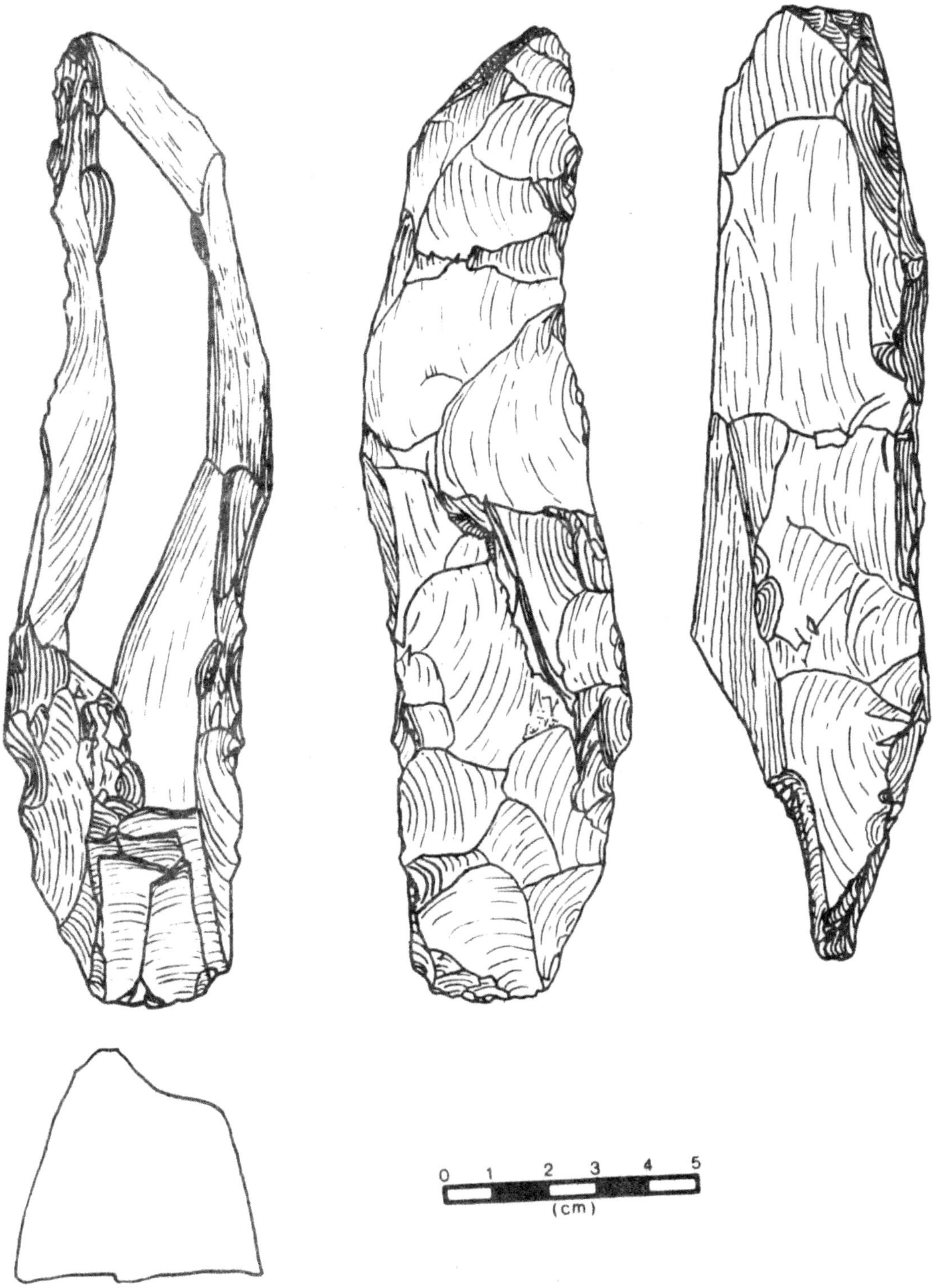


Figure 12.7 Triangular preform (apex up) made by bilateral trimming, with 'beak' reduced by removal of small blades to form cutting edge. Riverton D3 L-4

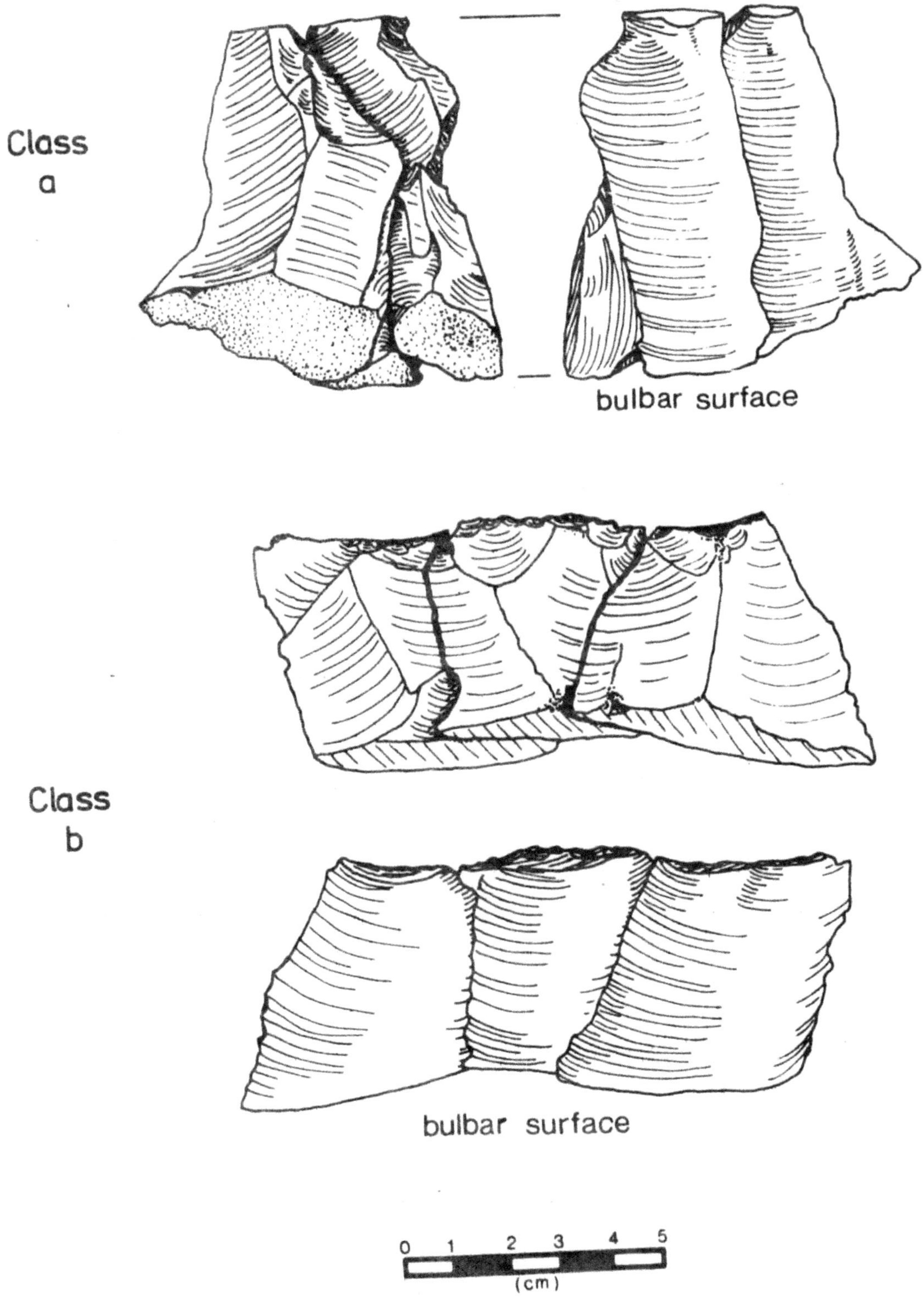


Figure 12.8 Primary (Class A) and secondary, trimming flakes (Class B). Only Class B flakes show overhang reduction scars.
Class A flakes: Riverton B2 L-3
Class B flakes: Riverton Z16 L-3A

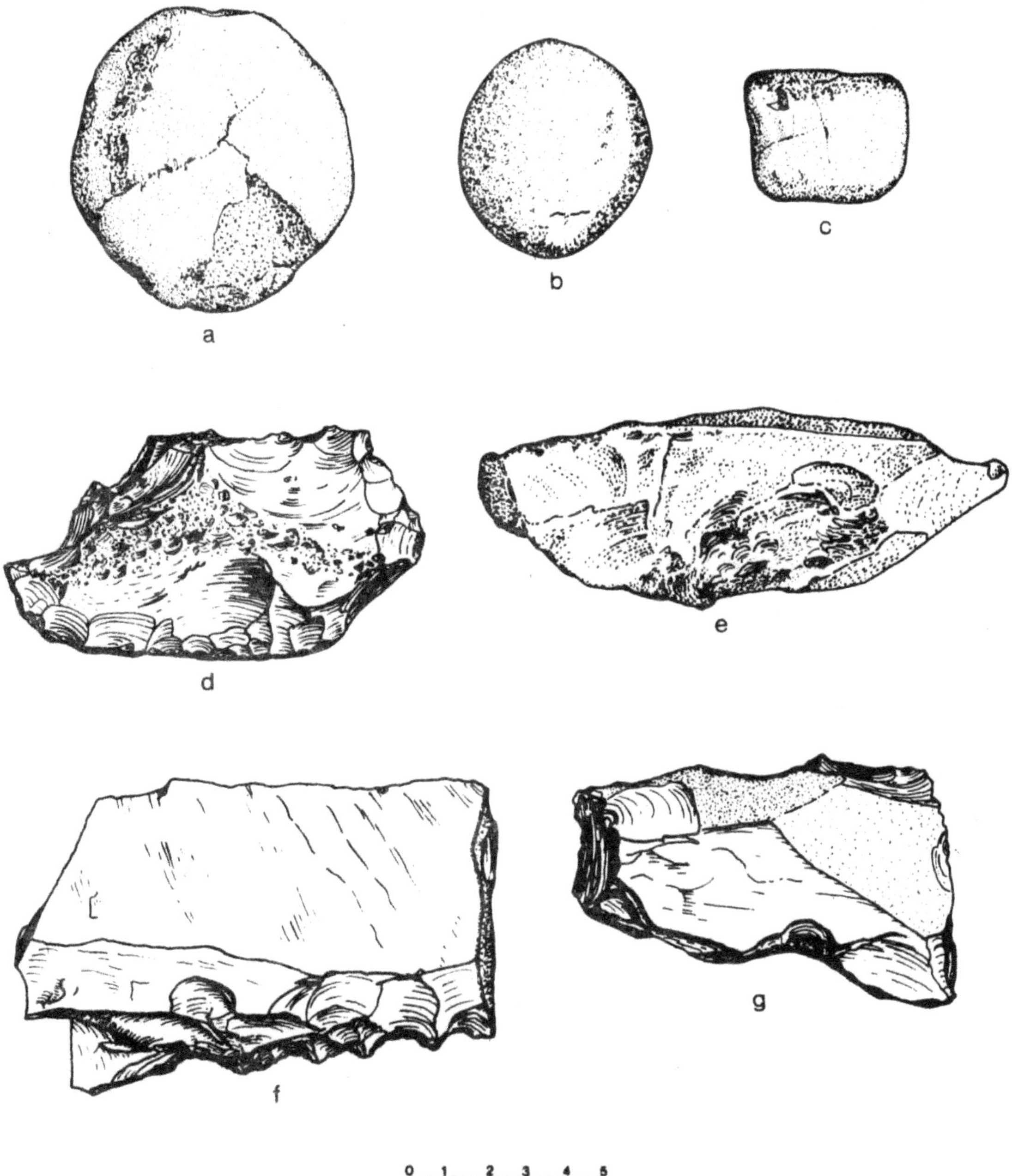


Figure 12.9 Hammers from the Riverton site. Top row: trimming hammers. Middle row: convex-edged hammers of uncertain function. Bottom: notched edge-reduction hammers.

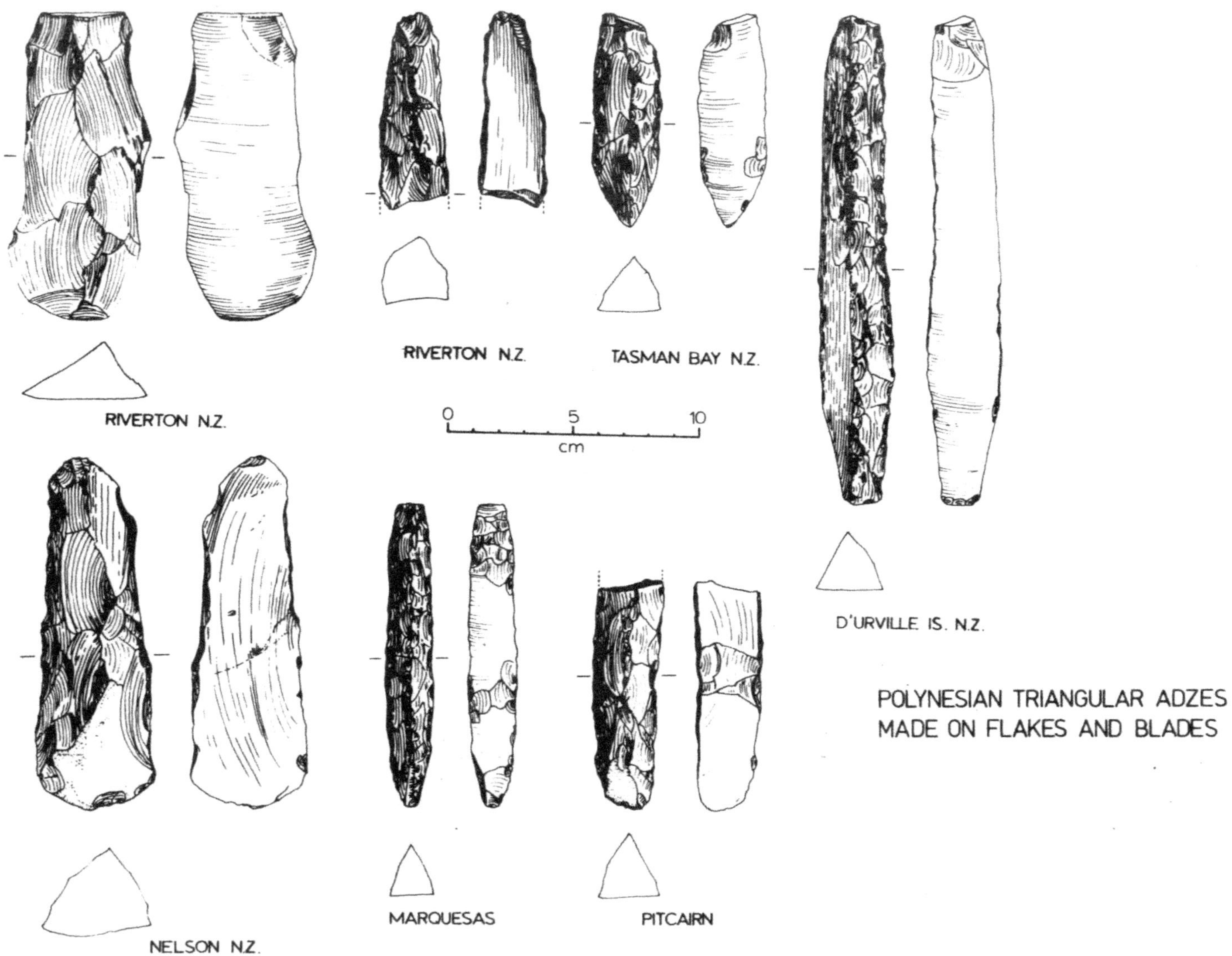


Figure 12.10 Polynesian triangular adzes made on flakes and blades. All show sinuous apex ridge of prepared core technique.

shows alternate sinuous flaking of the ridge identical to New Zealand specimens (Figure 12.10). These Marquesan blades are closely matched by triangular adze and chisel preforms from Pitcairn in the Otago Museum, some of which are made from blades and others from flakes.

It is apparent that the techniques of blade making were applied in several parts of East Polynesia in the adze and chisel manufacturing process, and that in a few locations blade tools were also made per se, generally at the beginning of the prehistoric sequence. More examples of triangular 'blade' adzes may exist among collections of preforms from other East or even West Polynesian areas. It is essential that their distribution and age are more fully documented.

In conclusion, pioneering studies of the technology of adze manufacture in Hawaii and New Zealand have already demonstrated that complex and probably highly diagnostic procedures are recognizable in the preforms and waste flakes of quarries and working floors. There is now growing evidence that technological factors in conjunction with the properties of the raw materials and the form in which they occur have played an important role in the modification of the Polynesian adze assemblages.

## REFERENCES

Best, S., 1977. The Maori adze: an explanation for change. Journal of the Polynesian Society 86(3): 307-337.

Duff, R., 1946. Native quarries of baked argillite. Records of the Canterbury Museum 5: 115-24.

Duff, R., 1959. Neolithic adzes of Eastern Polynesia pp. 121-147 In: Freeman, J. D. and W. R. Geddes (Eds.), Anthropology in the South Seas. Thomas Avery, New Plymouth.

Duff, R., 1970. Stone adzes of Southeast Asia. Canterbury Museum Bulletin 3.

Emory, K. P., 1968. East Polynesian relationships as revealed through adzes pp. 151-169. In: Yawata, I. and Y. H. Sinoto (Eds.), Prehistoric Culture in Oceania. Bishop Museum Press, Honolulu.

Gerard, B., 1976. Outillage sur éclat d'Eiao, îles Marquises. Journal de la Société des Océanistes 32: 98-102.

Green, R. C., 1969. Excavations at Va-I, 1963-64 pp. 114-137 In: Green, R. C. and J. M. Davidson (Eds.), Archaeology in Western Samoa. Vol. I. Auckland Institute and Museum Bulletin 6.

Green, R. C., 1971. Evidence for the Development of the Early Polynesian adze kit. New Zealand Archaeological Association Newsletter 14(1): 12-44.

Green, R. C., 1974a. Excavation of the prehistoric occupations of Su-Sa-3 pp. 108-154 In: Green, R. C. and J. M. Davidson (Eds.), Archaeology in Western Samoa. Vol. II Auckland Institute and Museum Bulletin 7.

Green, R. C., 1974b. A review of portable artifacts from Western Samoa. pp. 245-275 In: Green, R. C. and J. M. Davidson, (Eds.), Archaeology in Western Samoa. Vol. II Auckland Institute and Museum Bulletin 7.

Green, R. C. and Dessaint, A. Y., 1977. Measurements of Polynesian adzes: some comparative results for Samoa and New Zealand. Oceanic Prehistory Records 5.

Von Haast, J., 1880. Notes on an ancient manufactory of stone implements at the mouth of the Otokai Creek, Otago. Transactions of the New Zealand Institute 12: 150-3.

Huffadine, M. S., 1978. To leave no stone unturned: an examination of a lithic assemblage from Tiwai Point, Southland. Unpublished Research Essay. Anthropology Department, University of Otago.

Keyes, I. W., 1961. The age of the quarry sites on D'Urville Island. New Zealand Archaeological Association Newsletter 4: 30-2.

Keyes, I. W., 1975. The D'Urville Island-Nelson metasomatised rocks and their significance in New Zealand prehistory. Whakatane and District Historical Society Historical Review 23(1): 1-17.

Leach, B. F., 1969. Radiocarbon dates for the Oturehua Quarry Site, Central Otago (S134/1). New Zealand Archaeological Association Newsletter 12(1): 52.

Leach, H. M. and Leach, B. F., 1980. The Riverton Site: an Archaic adze manufactory in Western Southland, New Zealand. New Zealand Journal of Archaeology 2: 99-140.

McCoy, P. C., 1976. A note on Easter Island obsidian cores and blades. Journal of the Polynesian Society 85(3): 327-338.

McCoy, P. C., 1977. The Mauna Kea Adze Quarry project: a summary of the 1975 field investigations. Journal of the Polynesian Society 86(2): 223-244.

McCoy, P. C. and Gould, R. A., 1977. Alpine archaeology in Hawaii. Archaeology 30(4): 234-243.

Métraux, A., 1940. Ethnology of Easter Island. B.P. Bishop Museum Bulletin 160, Honolulu.

Moore, P. R., 1976. The Tahanga basalt: an important stone resource in North Island prehistory. Records of the Auckland Institute and Museum 13: 77-93.

Shaw, E., 1963. Maori quarry, Tahanga Hill, Opito. New Zealand Archaeological Association Newsletter 6: 34-6.

Skinner, H. D., 1913. An ancient Maori stone-quarry. Transactions of the New Zealand Institute 46: 324-9.

Skinner, H. D., 1974. Comparatively Speaking. Studies in Pacific material culture 1921-1972. University of Otago Press, Dunedin.

Thomson, J. A., 1918. Maori rock-quarries on D'Urville Island. New Zealand Journal of Science and Technology 1: 321-1.

Verin, P., 1969. L'Ancienne civilisation de Rurutu. Memoires O.R.S.T.O.M. 33.

Walls, J. Y., 1974. Argillite quarries of the Nelson mineral belt. New Zealand Archaeological Association Newsletter 17: 37-43.

# 13. VALUED GOOD OR VALUABLE STONE? CONSIDERATION OF SOME ASPECTS OF THE DISTRIBUTION OF GREENSTONE ARTEFACTS IN SOUTH-EASTERN AUSTRALIA

Isabel McBryde and Graham Harrison

In 1862 Victorian pioneer Isaac Batey was working on sheep stations in the Riverina, the flat stoneless plains between the Murray and Murrumbidgee Rivers in south-eastern Australia (see Figure 13.5). He stopped one day to talk with an Aboriginal drover from the Lachlan.

> .... Questioned as to how stone tomahawks were obtained ... my query ... elicited the reply 'from a hill down in the Melbourne country' ... I learnt from him that our aboriginals went inland carrying stone implements with other things. What they brought was exchanged with remoter tribes for what they produced, hence it appears that ... [they] ... have what I shall call the commercial instinct. From this Lachlan native it is clear that his people knew well where stone suitable for certain purposes was to be had in Victoria. Conjecturally his native place was 300 miles in a direct line from Melbourne and probably diorite with flints found their way hundreds of miles beyond where he was born (Batey, n.d.: 129-130).

Batey identified the stone sources referred to by his Lachlan informant with the quarried outcrops on Mt William and Mt Camel in the ranges north of Melbourne (Figure 13.1). His estimate of the extensive distribution of stone from these quarries in the central greenstone belt of Victoria was perceptive. It has an impressive dispersal array of some 700 km to the north-west and 550 to the west (see Figure 13.5).

In this paper we present the latest data on this distribution and examine certain aspects of that for two of the major quarries, to test hypotheses concerning the exchange processes which might be expressed in the distribution (McBryde, 1978). The investigation of the greenstone quarries is a long term project being conducted by one of us (I. McB.) combining archaeological and ethnographic research with petrological and geochemical source characterization and discrimination studies. It aims at understanding the archaeology of the quarries and their exploitation, at establishing the distribution of their products and interpreting this in terms of exchange, social interaction and relationships (see McBryde and Watchman 1976; McBryde 1978; McBryde 1979a; Watchman and Freeman 1979). The responsibilities of the two writers of this paper, in archaeological and statistical concerns are clear. The altered volcanic rocks of the greenstone belts (amphibole hornfels at Mt William and Mt Camel, andesite at Berrambool) provide ideal raw material, in their toughness, hardness and fine grain for ground-edge heavy duty tools. In the literature Mt William is reputed to be the major source for tomahawk stone in south-eastern Australia. However, it is just the largest and best known of a number of important quarry sites in the greenstone belts (Figures 13.1-13.4).

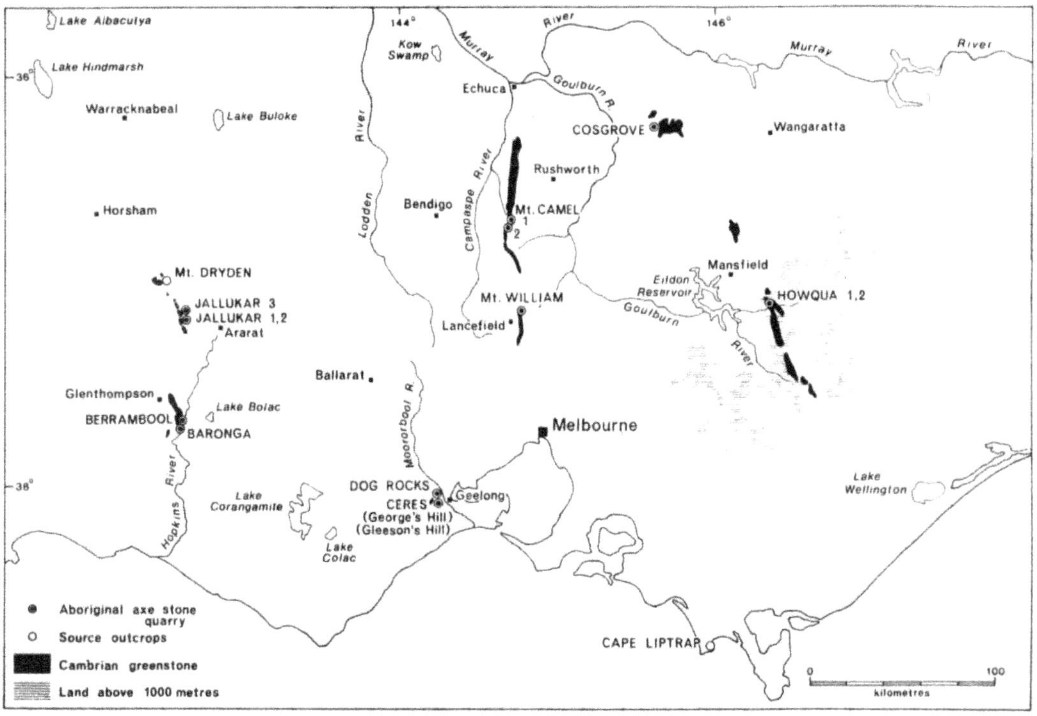


Figure 13.1 Quarry sites in the greenstone belts of south-eastern Australia.

Figure 13.2 The Mt William quarry site from the air. Arrows indicate the limits of quarrying activity, which extends for over a kilometre along this ridge on the northern slopes of Mt William.

Figure 13.3 A: Evidence for separation of activities is a feature of the Mt William site, with scree slopes of quarried waste around the worked outcrops and isolated flaking floors indicating the location of the areas where cores and preforms were shaped. This sequence can be seen down the slope in this photograph with a large flaking floor at base. B: On the top of the ridge at Mt William where the outcrop is buried there are strings of quarry pits where unweathered stone was mined.

Figure 13.4 Mt William quarry site; down slope of the exposed worked outcrops are dense concentrations of quarried waste and areas of flakes and worked pieces where core and preform manufacture took place.

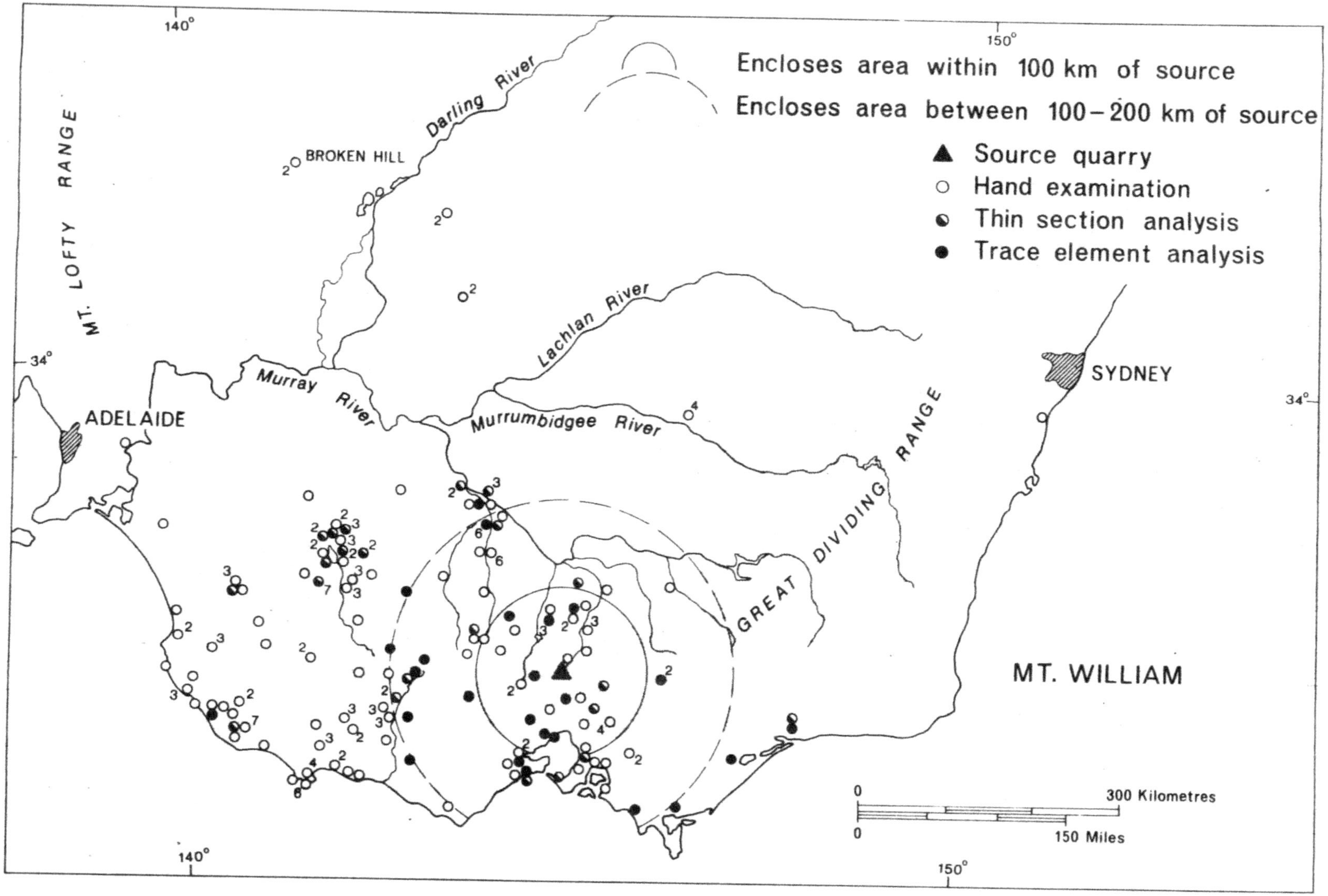


**Figure 13.5** Distribution of artefacts derived from the Mt William quarry site. The different symbols indicate different techniques used to make the source discrimination for specimens.

In the dispersal arrays and distance-decay curves drawn up for artefacts quarried from these various outcrops and published in an earlier paper (McBryde, 1978) there emerge intriguing variations between those of the central belt sites (Mt William and the Mt Camel quarries) and the others such as Berrambool in the west and Howqua in the east. This variation involves not only distance, but also its direction. Visual examination of the distribution plotted for Mt William artefacts and those from the Mt William/Mt Camel group[1] compared with those from Berrambool shows certain features. The distribution of artefacts derived from the three quarries in the central belt, particularly from Mt William, is extensive. It is comparable to the extent plotted for material from the quarries of Neolithic Britain such as the porcellanite sources at Tievebulliagh in Northern Ireland or the Langdale sites in Cumbria. Given the size of tribal territories in the nineteenth century in south-eastern Australia, and the variety of ecological and lithic resources it offered, such extent raises interesting questions. It is too great and the numbers involved too substantial to represent the seasonal movement of a population with casual loss of tools during any wide-ranging mobile annual subsistence régime. The distances might not have such interesting implications if recorded in the harsh lands of the arid interior where tribal territories were larger, group movements more frequent and such journeys longer. The numbers involved could also suggest some more regulated process than casual individual exchanges of goods.

The dispersal arrays are also non-random, and non-uniform. They display certain firm directional trends—to the north-west and west for the central belt quarries, to the south and west for the Berrambool site. These directional trends do not seem to correlate with factors of local geology nor with the availability of suitable hard rock. Mt William artefacts move in numbers into the Western District of Victoria rich in basalts of high quality and where the local greenstones are exposed along the Hopkins River and quarried at Berrambool and Baronga. The areas of concentration of artefacts from these quarries do not entirely coincide with areas of known high population density as reflected in either the nineteenth century records of group size and tribal territories, or the distribution of archaeological sites (see McBryde, 1978: 362-363; Lourandos, 1977).

Turning from the distribution maps to distance-decay curves for material from these quarries we find comparable variations between the sites, for

---

1 In this project the characterization of the quarry material and sourcing the artefact collections has posed complex petrological and geochemical problems (See Watchman in McBryde 1979a and Watchman and Freeman 1979). Even on trace element analysis certain parts of the Mt William and Mt Camel outcrops cannot be discriminated. So a substantial group of artefacts cannot yet be assigned to either source, but has been regarded as belonging to a separate Mt William/Mt Camel group until more detailed analyses enable them to be distinguished. At present some of this group are being studied by neutron activation analysis in England which may enable us to further discriminate. For this particular study the Mt William/Mt Camel group is examined together with that from Mt William for the size analyses, as it must include a considerable number of artefacts from that quarry.

example between Berrambool and Mt William or Mt Camel (see McBryde, 1978: 373). The flatter, elongated curves for Mt William and Mt Camel contrast with the short steep curve for the Berrambool site. For this site 60 per cent of the sample is located less than 100 km from source, none beyond 300 kilometres. This curve is reminiscent of those interpreted by Renfrew as characteristic of 'down-the-line' exchange, those of the central belt of his 'directional trade' with its concentration effects associated with redistribution centres (Renfrew, 1977: 85-87). Certainly the anomalies in these curves appear consistent, and may be located geographically. They seem to indicate some concentration point, or node, or secondary source in the Western District of Victoria (McBryde, 1978: 364-365, Figures 9 and 11). The ethnographic record describes several ceremonial or exchange centres for this region, such as that at Mt Noorat. However this convenient match must be tested further as the interpretation of distance-decay curves is difficult. Many different processes of exchange may be plotted to produce apparently similar curves of distance-decay (cf. Hodder, 1978: 156-163), while the poor resolution of most archaeological evidence provides an added dimension of uncertainty. However, granting these problems, the differences between the curves for the two groups of quarry sites do suggest the likelihood of differing exchange processes being involved in their distribution.

Another feature of the distributions as plotted, especially for the Mt William and Mt Camel material is their absence from certain regions, for example the south-east of Victoria (Gippsland) and the areas west of the Murray in South Australia. In earlier studies it seemed that the Wimmera/Mallee region of north-western Victoria was a third area into which the greenstone artefacts were not distributed. Though not as densely 'populated' as adjacent regions to north or south (perhaps predictable given the relative environmental richness of the three regions), central belt greenstones are found there in numbers (Compare McBryde, 1978: Figures 2 and 4 with Figures 13.5 and 13.6 in this paper). This revision of earlier conclusions followed field work in the Wimmera/Mallee and examination of several very large local artefact collections which expanded the evidence available in the collections of the three major state museums in Adelaide, Melbourne and Sydney, and the National Ethnographic Collection in Canberra. It was a salutary lesson in the problem of sampling for such a project and of the use of ethnohistorical evidence (see McBryde, 1979b). Berrambool artefacts, however, do not seem to have reached the Wimmera/Mallee, only 200 km to the north, which raises interesting questions about the direction of flow of Mt William and Mt Camel artefacts into the region if some social barrier excluded those from Berrambool in the south where all three groups seem to share dispersal patterns and presumably dispersal mechanisms. An earlier paper discussed the features of the greenstone distributions in relation to population densities and distribution, and in relation to the distribution of relevant lithic resources, and the outstanding qualities of the greenstone itself for tool-making (see McBryde, 1978). These one must admit as powerful determinants of the recorded distributions. The stone is of high quality, eminently suited to the manufacture of heavy duty tools subjected to high impact shock; it must be considered a 'high quality good'. Certainly all these factors are relevant, and necessary causal conditions, but none sufficient in itself, or even in combination to explain all the features of the distribution or of the distance-decay curves for the artefacts derived from the central belt greenstone quarries. Hence interpretations so far offered (for example in McBryde, 1978) have stressed the importance of social factors and patterns of group interaction in determining the particular features of the distribution, suggesting that the extent of dispersal of material from Mt William and Mt Camel, and the numbers involved, reflected distribution processes differing from those related purely to supply and demand and the spread of an economic technological necessity. Social barriers, traditional group alignments and hostilities may be invoked to explain the areas of non-penetration. Gippsland, for example, was occupied by a group of tribes known as the Kurnai, traditionally at enmity with the tribes of the Kulin group in whose territory the Mt William quarry is located. Similarly, it is reported that the tribes of the Adelaide plains were isolated from access to exchange goods from the east by their enemies on the lower Murray (Howchin, 1934: 23-24). Considering the features of dispersal, and of the distance—decay curves (with their marked differences between central belt sites and those of the western belt) we might posit differences in the distribution processes concerned. Given the distances and numbers involved these must reflect exchange processes or systems.

From examination of the distribution arrays and distance-decay curves (both directional and direction independent), and of the relevant ethnographic evidence, certain hypotheses concerning the distribution processes could be put forward. We could posit a supply and demand acquisition of the material from the Mt William site in direct barter by those of the Kulin group of allied tribes in central Victoria. Those wishing to acquire this stone must travel to the source and negotiate an exchange. The ethnographic records mention reed spears being brought from the Goulburn and Murray Rivers to exchange for stone (Smyth, 1878, I: 181 and II: 298-9; Howitt, 1904: 311-312). The value of the stone at source may be gauged in the barter of three pieces of Mt William stone for one possum skin cloak, itself a considerable labour investment in hunting, skin preparation, sewing and decoration, when the skins of many animals might be needed for one garment (Wright, 1979). Further, the Wudthaurung of the Geelong area sought Mt William stone even though their own territory contained excellent hard rock sources in the outcrops at Ceres and Dog Rocks (Figure 13.1) (Howitt, 1904: 312, 340-341). Was their need therefore of stone for subsistence or social functions? Perhaps it was for use as a valuable to offer as an exchange good at the meetings and ceremonial gatherings in the Western District which they attended bringing axe stone for exchange as Dawson records (Dawson, 1881: 3, 78). Access to the Mt William quarry was strictly limited, and its stone could be worked only by specialists/'professionals' in the sense of those with the requisite social status and kin affiliations (the 'owners' of the site) (Howitt, 1904: 72, 311-312). So the quarry acquires certain characteristics of a ceremonial, sacred place. Similar associations and restrictions of access and mining applied to the Nillipidji quartzite blade quarry in Arnhem Land recorded by Thomson in 1935 (Thomson, 1949: 65, 87). In recent field work Morphy found this site (known locally as Rarramalakmi) the subject of considerable mythological associations and reverence, with strict controls of access, quarrying and the distribution of the quarried blades (H. Morphy, pers. comm.).

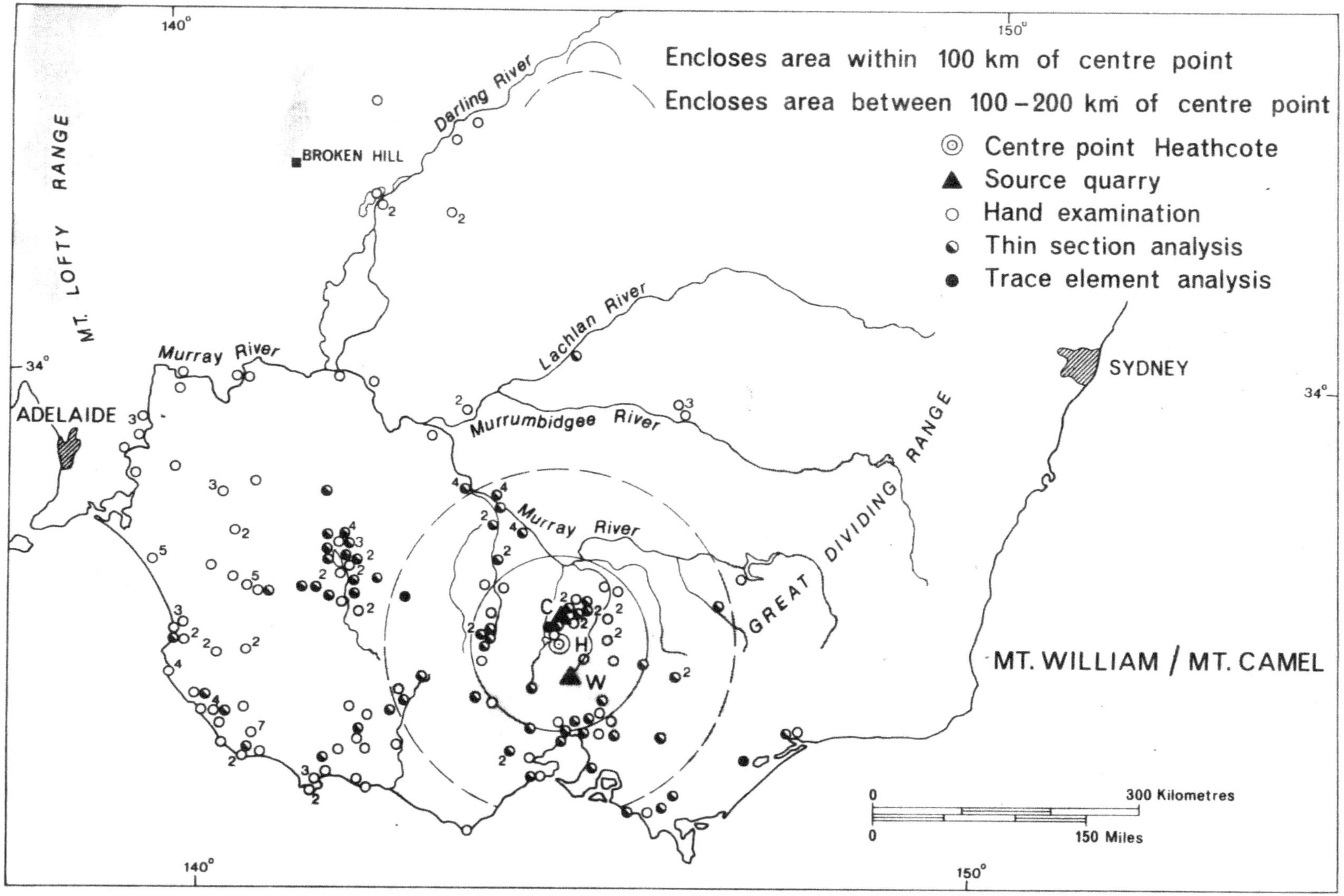


Figure 13.6 **Distribution of artefacts belonging to the Mt William/Mt Camel group, quarried from these sites in the central greenstone belt (for definition of this group see p. 189).**

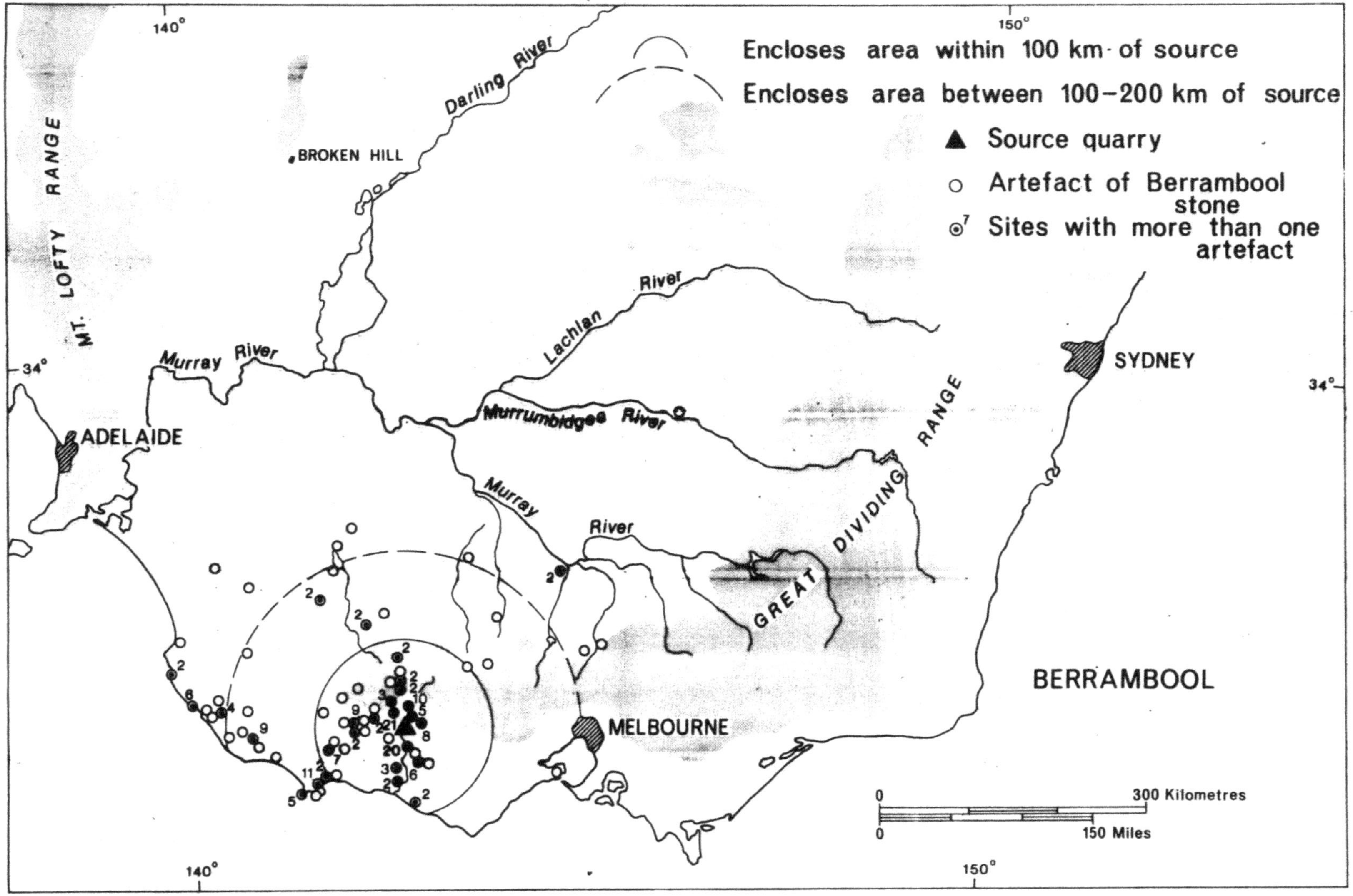


Figure 13.7 Distribution of artefacts quarried from the Berrambool site on the Hopkins River in the Western District of Victoria.

So, though the Mt William site could in no way be considered a commercial, industrial enterprise producing for profit, the restrictions on access and use show some features which might be considered characteristic of such. Within the Kulin territory therefore we can see the site as an important resource belonging to one clan of the Wurundjeri (part of the Woiwurrung linguistic group), its products seen as valuable for the manufacture of essential heavy duty tools and acquired by direct barter at the source. Further distribution presumably took place beyond the quarry by direct individual exchange.

Outside the Kulin territory one could postulate that the distribution process changes. The artefacts could be channelled through the nodes of the ceremonial centres of the Western District (for example Mt Noorat) which also served as exchange centres, into the exchange networks of the populations of areas further west and north (for location of these centres see the map in McBryde, 1978: Figure 11). The motive for exchange here is more likely to be social and ceremonial than economic or utilitarian; to fulfil, cement or create social ties or ritual obligations. For such purposes a high value good is required. The direction of the distribution thus initiated will be determined by the exchange networks and kin affiliations of the individuals involved rather than by any bulk transfer of goods at 'trading centres'.

The distribution of central belt greenstones outside the territories of the Kulin group of tribes into the south-western parts of the sample area could derive from the ceremonial centres of western Victoria, serving as secondary sources or re-distribution centres. The importance of the stone could change from that of a valuable utilitarian object to that of a valued good of certain prestige derived from a known source. Here one may invoke the analogy of the Nillipidji spear head blades whose prestige (marr) (derived from their source and the renown of the craftsmen producing them) increased with distance from the sourcequarry until they acquired almost a spiritual quality (for this summary of Thomson's views on the Nillipidji blades I am indebted to Mrs Thomson - D. Thomson, pers. comm.).

Distribution array and distance-decay curves for numbers of the artefacts from the Berrambool quarry suggest the full exploitation of a valuable local resource and the movement of its useful products within a limited area. The majority cluster within 100 km of the source (Figure 13.7). One could posit distribution and acquisition of this material in ways similar to those proposed for the Mt William material within Kulin territories. There is no historical information on the use of this quarry comparable to that available for the Mt William site. Beyond the Western District of Victoria Berrambool artefacts are found to the south and west on sites up to 300 km from the source, but very rarely to the east or north. They are not found on sites in the Wimmera/Mallee desert region between the Western District and the Murray though central belt greenstones are. Perhaps another social barrier is reflected in this archaeological pattern, the isolation of two adjacent populations, and with it a clue to the directions from which the central belt greenstones entered the Wimmera/Mallee—not from the south but from the Murray lands to the north. These aspects must be tested in later detailed and regionally oriented analyses, with careful control for effective rather than linear distances. The analyses which we will present here would not be appropriate to problems of this scale.

In this paper we present the first stage of an examination of the attributes and distributions of the greenstone artefacts in south-eastern Australia as preliminaries to interpretation in terms of processes of exchange. These in turn of course have implications for economic and social interaction and relationships, areas often elusive in archaeological studies. Given the hypotheses on the exchange processes involved in the spread of Mt William and Berrambool artefacts outlined above, which were derived from the ethnographic information and study of the distribution arrays and distance-decay curves for the sites, we plan to test these against certain models of dispersal. First we must accept that there is a significantly different distribution of greenstone and non-greenstone artefacts across the sample area, for both groups are not found uniformly distributed across the landscape but vary in different ways, with the non-greenstones conforming to predictable patterns based on expected areas of high population densities while the greenstones form an eccentric group within this pattern trending to the south and north west. They are absent from certain environmentally favoured regions such as Gippsland. Granting this we shall then examine two groups within the greenstone, the Mt William (plus the Mt William/Mt Camel group - see p. 189) and the Berrambool group. If the previously accepted conclusion that the dispersal of these two groups of artefacts is different be confirmed, we wish to test whether this difference could reflect different processes of exchange which may in turn reflect the perceived value of the item being distributed. The proposition is that the Mt William stone in its local Kulin territory is seen as a high quality item for technology and subsistence tools requiring hardness and resistance to impact shock. Beyond that territory (especially to the south-west where it intrudes into areas with local greenstone quarries and widespread basalt outcrops) it is distributed not so much as valuable stone as valued good. It could be perceived as a high value item with prestige, so useful in the social or ritual context of reciprocity. Analogies from Australian ethnography may be cited in support of this hypothesis. The quartzite blades from the Nillipidji (Rarramalakmi) quarry often circulate over great distance in exchange systems still in their original paper-bark bundles, never unwrapped or hafted for use; presumably their value is greater as a potential exchange good in the social arena than as spear head in the subsistence arena. Locally available quartzites could supply this need. From New Guinea there are analogies in the distinctions between the work axe and the ceremonial axe, and the social or ceremonial functions of axes and their subsistence functions. These distinctions are clearly perceived and important, yet are also two ends of a fluid continuum capable of considerable internal variation (Hughes, 1977: 142, 160-167; Strathern 1965: 185; Strathern 1969: 320-323).

The hypothesis to be tested relating to the Berrambool artefacts proposes that these are acquired and distributed as utilitarian items valuable for technology and subsistence, and are widely used and widespread within the region of the source. Of the sample 60 per cent are discarded or lost, and so enter the archaeological record, within 100 km of the quarry. Certainly some examples reach more distant localities, but only in small numbers, such as are consistent with casual exchanges between individuals and need not counter the broad hypotheses outlined above.

The hypotheses outlined could be tested at a number of levels and in many ways. We intend following through with a variety of tests and varying approaches using differing attributes of the collections and differing controls of spatial variation. We intend to look at effective distances rather than linear distance, and to model against topographic features rather than assuming a uniform landscape as we shall have to do in this particular exercise. In this study we begin the testing process at a very basic level to open up the question and explore its parameters. We shall look at the variability in weight or mass as an expression of size for the artefacts in relationtto their distance from quarry source. The attribute we are using in this study is size as reflected in 'mass' (calculated as product of length, width and thickness) taking as our sample those artefacts which are complete specimens and for which we have all these attributes recorded. So the sample size for this analysis is limited compared with that expressed in the distribution maps (Figures 13.5-13.7). If the perceived value of the stone from these two sources varies then it should be reflected in variations in size/'mass' across distance. Other attributes may of course be relevant here, and may ultimately prove better indicators of this aspect, but for analysis to open up the problem we will consider size alone, as expressed in artefact 'mass'. One might predict that the valued good as against the valuable stone will travel further and show less steep decline in numbers against distance. As regards variations in size over space two possibilities might emerge—one that the valued good will maintain its size over distance, because it is retained for its social rather than its technological importance. The analogies of the Nillipidji spear head blades and the careful curation of certain New Guinea ceremonial or bride-wealth axes are relevant here (Hughes, 1977: 142; Strathern, 1969: 320-323). Conversely, the value of these goods might be reflected in their retention in use over long periods of time, with much re-sharpening and re-use, when less valued but equally useful stone tools would have been discarded at an earlier stage and larger size. This second proposition assumes that the valued good is also serving as a work tool, not reserved primarily for its social or ceremonial functions as is assumed in the first proposition. For the 'valuable stone' group one might expect strong concentrations in the vicinity of the source, that discard will take place fairly rapidly and so of relatively large items, still capable of re-sharpening and re-use. However, if the stone is readily available, easily acquired and has little exchange value then artefacts (particularly if broken or damaged) will be discarded at an early stage in their potential tool lives to be replaced by newly quarried pieces.

This hypothesis assumes ready availability of suitable stone for the manufacture of chopping/cutting tools, and should be tested against data from such regions. It may not be appropriate to areas where this does not exist, for example the Wimmera/Mallee sand desert area with no local hard rock outcrops. Here broken items might be retained for re-use in a variety of possible forms as all hard rock will be of value. Such areas must be regarded as special cases requiring individual study.

To test the hypotheses set out we intend to examine the relationships between size (as 'mass') and distance from source for the ground edge hatchet heads from the sample area as defined below (see p. 197). The sample will include both greenstones and non-greentones with the latter as a control. Of the greenstones, material from the Berrambool quarry will be used and that from the Mt William site (including the Mt William/Mt Camel group for reasons already discussed). These two samples are selected for the interesting variation already discernible in their distributions and distance-decay curves analysed in terms of numbers of artefacts against distance. The Mt Camel sites are omitted from the analysis. The distributions of Mt William and Mt Camel artefacts overlap in space, but are not entirely co-incident. Until these variations are more fully understood it seems better to exclude Mt Camel artefacts from the present exercise. In comparing Berrambool and Mt William one may be making assumptions about contemporaneity which are not warranted. The little evidence there is from archaeological excavation on dated Berrambool artefacts suggests use in the last millennium (H. Lourandos pers. comm.; Coutts and Witter, 1977: 67-68) while ethnographic records date the final use of the Mt William quarries to the early nineteenth century. However, for testing the present hypothesis on the processes involved in distribution from either quarry, contemporaneity of exploitation is not essential. There are assumptions that local stone suitable for making impact tools be available in the western parts of the sample area, but these need not involve use of the Berrambool quarry given the local basalt resources and the other greenstone outcrops along the Hopkins River.

Earlier discussions (McBryde and Watchman, 1976; McBryde, 1978; 1979a) were based on study of the internal evidence of the distribution of greenstone artefacts alone. For this analysis we include data on a sample of non-greenstone artefacts drawn from the same museum and private collections as the greenstone artefacts (for comments on the sampling problems for this investigation see McBryde, 1978 and 1979a). This addition to the sample will provide valuable comparative evidence on non-petrographic attributes and on distributions. It will establish needed base controls from which to understand the distributions and attributes of the greenstone artefacts. The total sample numbers over 3,000, drawn from the major state museum collections, local museums and large private collections. The sample area covers south-eastern Australia from $31^{\circ}$S to $39^{\circ}$S and from $138^{\circ}$E to $151^{\circ}$ 30"E. The distribution maps given here show the dispersal array over this sample area for the full artefact sample for the quarries used in this analysis. They up-date those previously published (McBryde and Watchman, 1976 and McBryde, 1978). They show all artefacts assigned to the sources selected for this study and the basis of source discrimination. In any investigation of this kind using museum collections sampling poses problems. Collector bias will be present, conditioning both the areas represented in the collections and the range of artefacts represented (in terms of typological categories and also in terms of lithologies given the handsome appearance of the greenstone specimens). The revision of earlier conclusions on the spread of greenstone artefacts into the Wimmera/ Mallee region reflects and illustrates these difficulties (see McBryde, 1979b: 145-146). The sampling problems associated with the project cannot be denied, but are shared by most archaeological studies. It must berecognised that few archaeological samples would meet the strict statistical requirements for 'sampling'. This only becomes a problem if we are trying to argue about populations. By being careful about the attributes considered these problems are avoidable.

Unfortunately archaeological 'samples' are, almost by definition, inadequate. The archaeologist is a specialist in making sense out of relict unrepresentative information. The survival characteristics of evidence of past activities are determined by durability not sample sufficiency and the collection of more and more information cannot 'improve' the sample or increase its 'statistical validity' (words almost without meaning in archaeology ) (Groube, 1978: 41)

Within the collections broad categories have been distinguished on typological and technological criteria. The sample includes both groups associated with the manufacture of ground edge tools—flakes, worked pieces, cores and flaked 'preforms' and complete ground edge tools. Detailed attribute records are kept for specimens as the basis for studies of the relationships between formal stylistic features, functional attributes and constraints and those imposed by the qualities of the raw material. This work is in progress and will test for correlations between features and quarry sources such as emerged in Hughes' analysis of New Guinea axes (Hughes, 1977: 143, 154, 160-168), and has been noted by other workers (Bulmer, 1964: 248 and Lampert, 1972). Comparative control will come from analysis of the non-greenstone components. The major category amongst the ground edge designed tools is that of the hatchet head, the blade either formed of an unmodified pebble or prepared by flaking or hammer-dressing with grinding confined to the actual cutting edge and its vicinity (see Figures 13.8 and 13.9). The term hatchet head rather than the more usual archaeological designation of 'axe head' follows the usage of nineteenth century observers and the arguments of Dickson (Dickson, 1978: 34-39). It emphasises that the stone element is the head of a short handled general purpose chopping/cutting tool used in one hand, not the equivalent of the modern long-handled axe swung with two hands. Within this category there is considerable variation in stylistic and technological features which will be examined in later analyses. From the collections two other categories emerge. There are ground edge flakes, small in size compared with the hatchet heads, relatively narrow and with thin, low-angled cutting edges. These are hardly functional variants of the first category, and presumably served as chisel bits or knives. The third category comprise large tools, in formal shape similar to the hatchet heads and often grooved for hafting, but they bear heavy battering on their working edges which has removed any grinding which might have existed. From the evidence for use we may presume them wedges or hammers rather than incomplete hatchet heads.

For the purposes of the present analyses of size ('mass') variation over distance we have included only the hatchet heads. This should improve control as lessening variation within the collection due to other factors. Inclusion of the other two categories which seem to represent functional classes within the collection between which weight might well be a discriminating attribute could mask any emerging trends due to variation over distance. Parallel studies of these categories as a test case will be made later and could prove a useful control.

For this preliminary analysis we shall be examining the distribution across the sample area using a lattice of rectangular quadrats based on the 1:250,000 map sheets for south-eastern Australia. These are large quadrats and so do not give any tight control over spatial variation. Their use as a measure of distance in this case (using quadrat centroids) could lead to minor distortions arising from the conversion of latitude and longitude measures to distance over the ground at these latitudes. However, for the present exercise this distortion may be accepted, particularly as it is no greater than that imposed by the practice of using quadrat centroids as the measure of distance. We see little justification for minimizing small errors, whilst ignoring larger ones.

For this examination we have confined the artefact sample to that group in the collection defined as hatchet heads (see p. 198), a functional class admittedly but separable from the other groupings in the collection on formal and technological attributes. Of these we include all complete specimens for which the three linear dimensions of length, width and thickness are recorded. The rejection of incomplete, broken specimens requires little discussion. The choice of 'mass' rather than weight as the comparative measure of size does need justification. Weight would have been a convenient measure, especially as it may have functional relevance for this type of artefact (though see below, p. 203). However, in recording the collections there were often circumstances in which it was possible to record linear dimensions, but not to weigh artefacts. The increased sample size was thought to more than compensate for the approximation to mass. A 'mass surrogate' has been achieved by calculations approximately proportional to volume (by multiplying the linear measurements). The 'unit' used in the tables and figures is approximately equivalent to a cubic centimetre as metric measurements were made of the artefacts. Variations from weight as a measure of size due to difference in the density of the rock types involved would be minimal for the two greenstone samples but could be greater for the diverse 'non-greenstone' group. However cross-checks with similar tests for weight measurements on the smaller sample available suggest no contradictions. The samples used for this analysis are shown in Table 13.1.

Table 13.1

Sample of hatchet heads used for analysis

| Quarry source | Total sample | Number for which 'mass' calculable | Mean 'mass' units S.E. |
|---|---|---|---|
| Mt William | 490 | 324 (66%) | 196 5.2 |
| Berrambool | 180 | 108 (60%) | 409 21.5 |
| Non-greenstone | 1845 | 572 (31%) | 317 9.7 |

The distribution of 'mass' in units for these three samples is presented in Figure 13.10, a histogram showing numbers at intervals of 60 units.

From examination of this histogram a number of general points may be made. Firstly, there is confirmation of the visual impression given by the artefacts themselves (see Figures 13.8, 13.9) that the surviving Mt William tools of this category tend to be smaller than both the Berrambool and the non-greenstone examples. This may reflect a number of factors, some technological, some possibly relating to perceptions of the value of the stone or to its inherent qualities such as its toughness. This variation between the three samples must be further investigated when looking at possible relationships

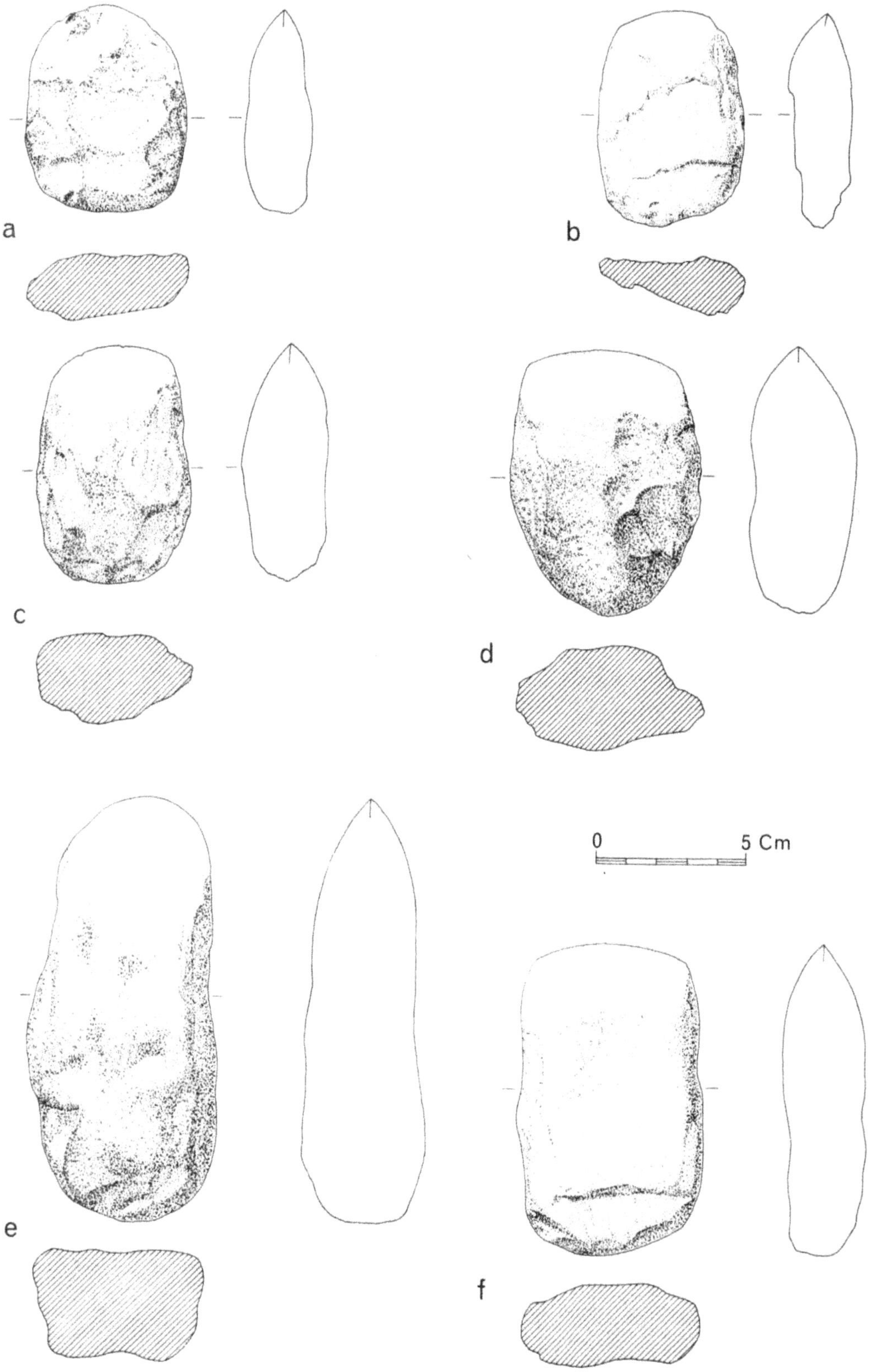


Figure 13.8 Hatchet heads of greenstone (amphibole hornfels) from the Mt William quarry (source determined on trace element analysis). These artefacts come from sites at Kerang in northern Victoria (c - no. 297; d - no. 205; f - no. 196), Swan Hill on the Murray (a - no. 166), Lake Bolac in the Western District of Victoria (b - no. 485) and Millicent in South Australia (e - no. 258). All these artefacts come from the Wishart and Mitchell collections of the National Ethnographic Collection, Institute of Anatomy, Canberra.

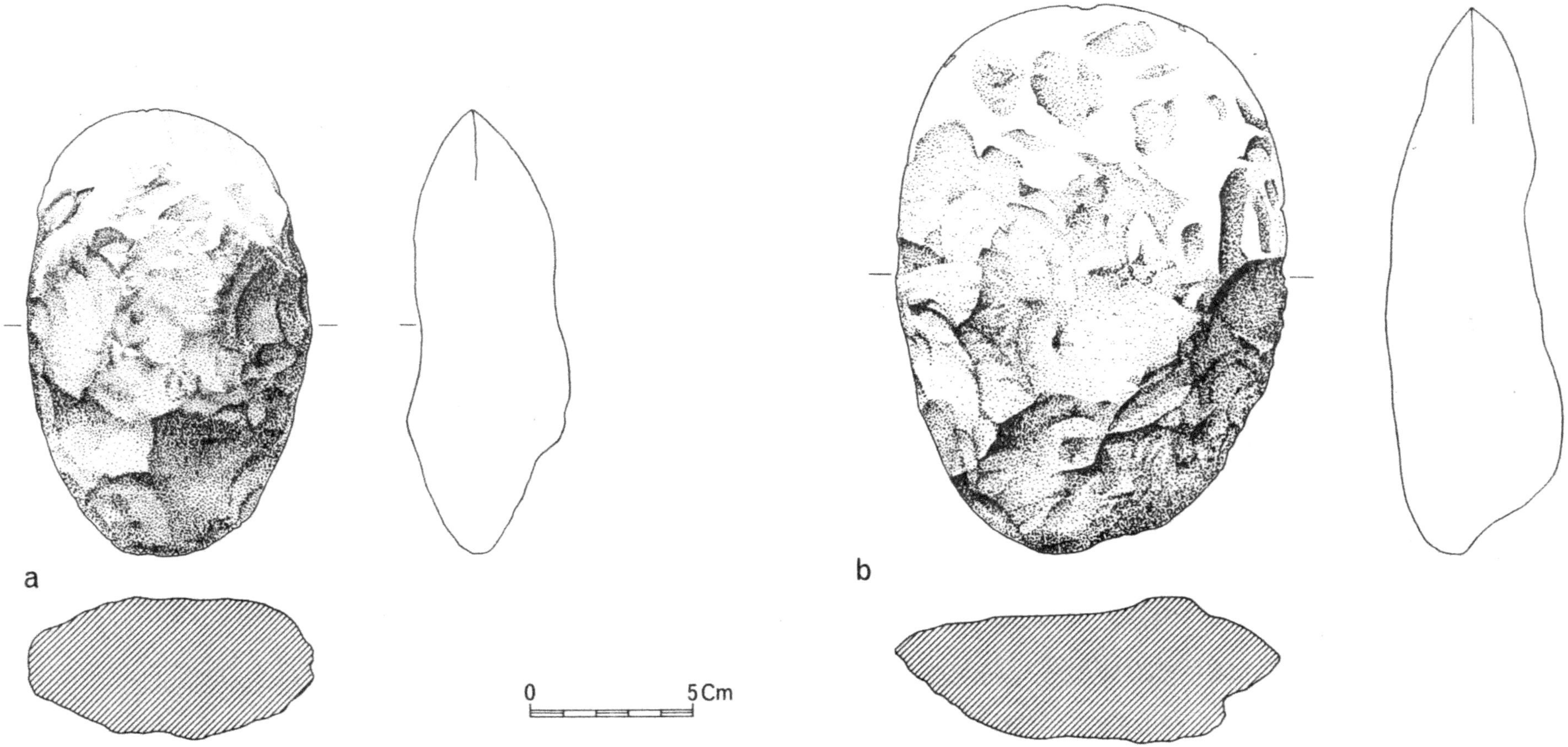


Figure 13.9 Hatchet heads of greenstone (andesite) from the Berrambool quarry site. Both come from sites on the coast west of the mouth of the Hopkins River, a (no. 399) from Narrawong and b (no. 426) from Portland. They were collected by S. R. Mitchell and form part of the National Ethnographic Collection, Institute of Anatomy, Canberra.

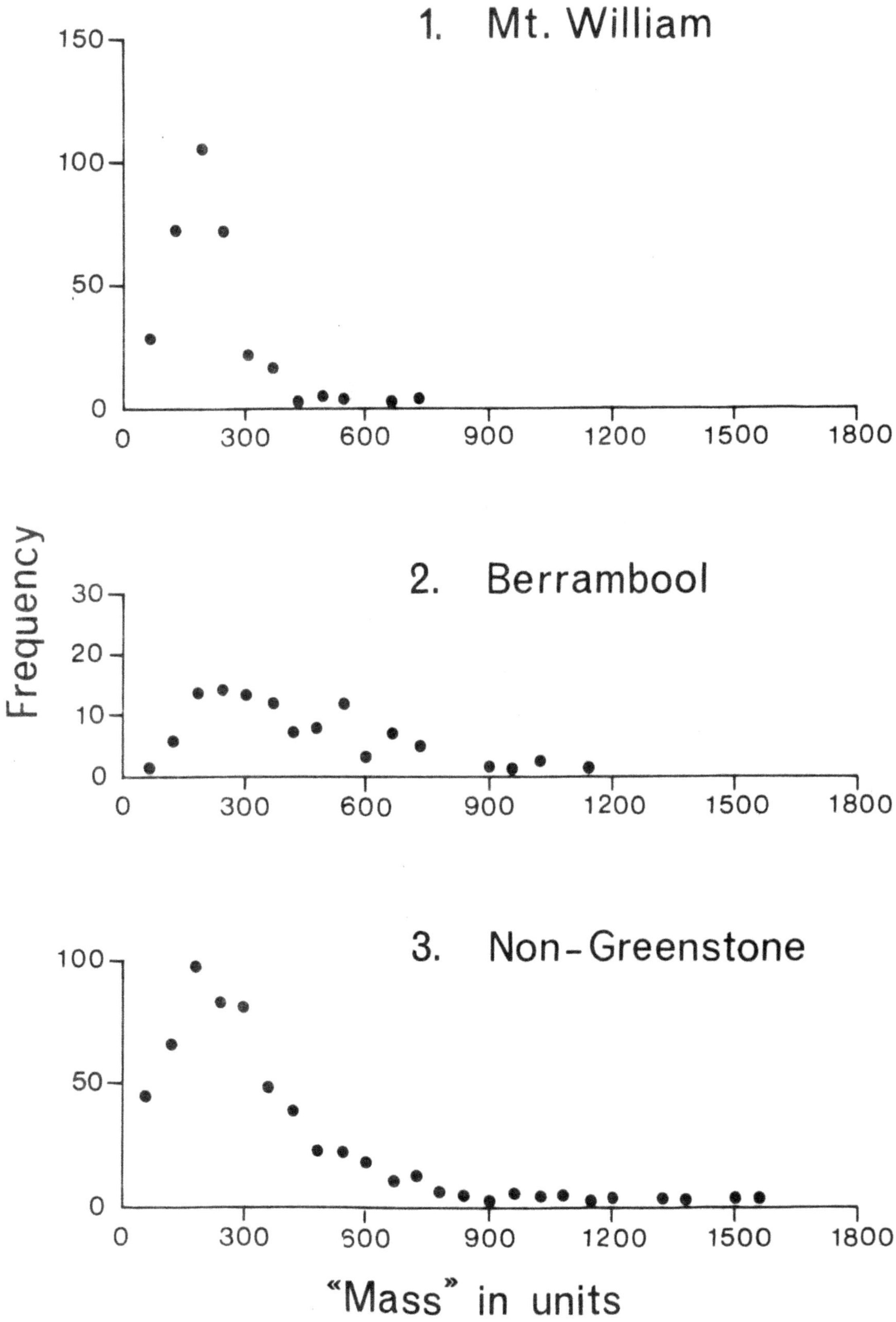


Figure 13.10 Distribution of 'mass' for the three samples of hatchet heads.

between artefact attributes and raw material. It need not indicate any inconsistency with the comments made above on the importance of weight as an attribute of the hatchet head. Dickson's studies of collections and his experiments have shown that within certain broad ranges of preferred weights for hatchet heads variability can and does occur. Variability also occurs in the length of the handle. The two elements of the composite tool may be adjusted to achieve the same effective use of energy in the impact tool (Dickson, 1978: 211, 218-219, 227, 235-246).

From Figure 13.10 it appears that the distribution of 'mass' for the non-greenstone artefacts is as expected, with skewness of a degree suggesting that weight or 'mass' was not the only variable involved in establishing the ultimate dimensions of the artefact. The sample from Berrambool (Figure 13.10) follows a rather similar pattern, though there may be some irregularity at about 600 units. The significance of this is yet to be established, and one cannot discount the effect of small sample size on this particular distribution. The Mt William distribution does not appear to display this pattern (Figure 13.10); possibly it can be divided into two parts, one (up to 300 units) with apparent symmetry suggesting that weight or 'mass' was the important criterion. However, the other part of the distribution represents a component of larger artefacts. This duality may well be relevant to the hypothesis outlined above involving two distribution processes in the spread of the Mt William material and two perceptions of its 'value'.

Figure 13.10 gives a simple presentation of the distribution of 'mass' in the samples from the three sources. It now remains to examine the distribution of 'mass' in samples as against distance from the source, initially with a direction independent analysis. The assumption that artefacts in a particular raw material will become fewer in number and decrease in weight or size over distance from their source is a reasonable one (cf. observations of New Guinea axes, that on the whole these are fewer in number, smaller in size and of poorer quality the greater the distance from source—Hughes, 1977: 134, 174; Strathern, 1969: 327; Chappell 1966: 113). Similar comments are made about the obsidians of the eastern Mediterranean. The assumption that this distance will be direction independent, that the distribution will be uniform across the landscape is less reasonable, given the likely impact of topographic or ecological factors, or of possible social ones. Even visual examination of the distribution arrays strongly suggests that some control for direction must be given to later analyses before any real understanding will emerge of the relationships between size and distance or of the determinants of the distribution itself. However many workers present direction independent studies, and as a preliminary stage in the analysis they are appropriate. A common method of assessing relationships between size and distance in such situations is to examine correlations between the two, to establish the ways in which size varies over distance and to calculate the slope of the line of best fit between the two, to give some idea of the best guess for the rate of change in size with distance. Table 13.2 presents the results of such analyses for the three samples in this exercise, relating 'mass' to the two quarry sources. Correlations and their associated significances are given.

Table 13.2

| Sample | Distance from quarry source Mt William | Berrambool | Number |
|---|---|---|---|
| Mt William artefacts | R = 0.07 p = 0.12 | R = 0.01 p = 0.42 | 320 |
| Berrambool artefacts | R = 0.17 p = 0.04 | R = 0.21 p = 0.01 | 106 |
| Non-greenstone artefacts | R = -0.04 p = 0.19 | R = -0.06 p = 0.07 | 541 |

Note: The small differences in total number in this table from those in Table 13.1 may be accounted for by those specimens whose position within the sample grid is uncertain.

Only the Berrambool sample shows any support for a hypothesis of linear change of size with distance from source. Large specimens are found furthest away. Satisfyingly such correlations as exist are stronger for these artefacts when the distance is measured from Berrambool than from Mt William, and the reverse is true for the Mt Williams artefacts. Such correlation as exists for the non-greenstone artefacts is stronger for measurements from Berrambool than from Mt William. Furthermore such difference as there is shows smaller non-greenstone artefacts further away from the two quarry sites.

The non-greenstone sample raises several interesting methodological points. Analysis of the non-greenstone in two categories, those in the north-eastern part of the grid (which represents southern New England and the Liverpool Plains) and the remainder shows that they are most unlikely to come from similar populations. The variance of the former is significantly less than the variance for the remainder of the sample ($F - 1.94 \quad p < 0.01$). The reason for choosing the split is that the north-east of New South Wales is an area into which greenstone artefacts derived from the Victorian quarries do not penetrate. So although it is within the geographic area covered by our model it is unlikely to be relevant to our discussions of greenstones or the arguments presented here. It may be necessary in future analyses to omit these areas but it must also raise questions about the homogeneity of regions within the remainder of the sample. The situation in the north-eastern section of our sample area provides interesting problems for later investigation. It may represent a cultural or stylistic boundary to be explained.

Although we have discarded notions of similarity on the basis of a variance ratio test it is worth remarking that the point estimates for means show a slightly smaller hatchet in the north-eastern New South Wales examples. This may be reflected in the little difference there is in the correlation coefficients.

Returning to the Berrambool sample, what is to be made of the fact that larger hatchets are being found further from the source? This is shown in the scatter diagram (Figure 13.11) which has plotted 'mass' against distance from quarry source. This implies that artefacts discarded or lost are entering the archaeological record at relatively large sizes, not that they are growing. Obviously stone tools cannot increase in absolute size. It appears then that from this analysis we may state the Berrambool artefacts of this category are larger in size when discarded at locations distant from the quarry, but as these distances are not extensive it need suggest little more than discard at a relatively early stage in artefact 'life history' and scant concern on the part of the maker or user to conserve the material from which the tool was manufactured. From the Mt William scatter diagram (Figure 13.11) it can be seen that in contrast these tools are recovered from locations at greater distances from the quarry, but do not follow any consistent pattern of change in weight with distance. This does not mean that there is none, for the diagram has hints of curvilinear relationship, also of a division of the trends in the collection at distances about 400 km from the source. These raise interesting questions for further study. One might also wish to examine the trends for disconformity at distances of 250 to 300 km from the source, which would represent the boundaries of the Kulin territories. Of the greenstone samples neither distribution seems random, nor to be explicable by a simplistic direction independent model of which the present coarse-grained study has made use. Yet it does indicate results requring closer examination, and results potentially relevant to understanding the processes of distribution for these materials. More controlled techniques of spatial analysis will be used in later studies, especially one more requiring closer examination, and results here (for general criticism of the quadrat analysis see Hodder and Orton, 1976: 33-38, they emphasise that it is a technique easily manipulated and hence a poor analytical tool). Control of direction, and of effective as well as linear distance, and modelling for likely determinants in topography, ecology, lines of communication, or social factors should allow us to examine these variations further, establishing their exact nature and their possible determinants. Experiment with the use of points rather than quadrat centroids as the measure of distance may prove useful. Studies of the changing proportions of greenstone as against non-greenstone artefacts for key sites and regions will allow an independent test of the same hypotheses using different approaches.

The results obtained in these preliminary studies are not at variance with hypotheses of differing exchange processes being involved and differing values being accorded the goods distributed, thus affecting their life histories as artefacts and the relative sizes at which they are discarded and so enter the archaeological record. The correlation seems to vary between the two quarries with a hint of retention of size for the Mt William artefacts at considerable distances from the source according with the hypothesis of the 'valued good'. The Mt William artefacts seem to maintain size more uniformly over distance than the other samples examined. There was not the degree of variation expected if the hypothesis of 'valuable stone' within the home Kulin territories, 'valued good' beyond them were to be supported. However, a certain curvilinear trend does emerge from the scatter diagram (Figure 13.11) which would be consistent with this hypothesis. Other techniques may enable us to understand this trend more clearly and relate it to possible explanations. Certainly these

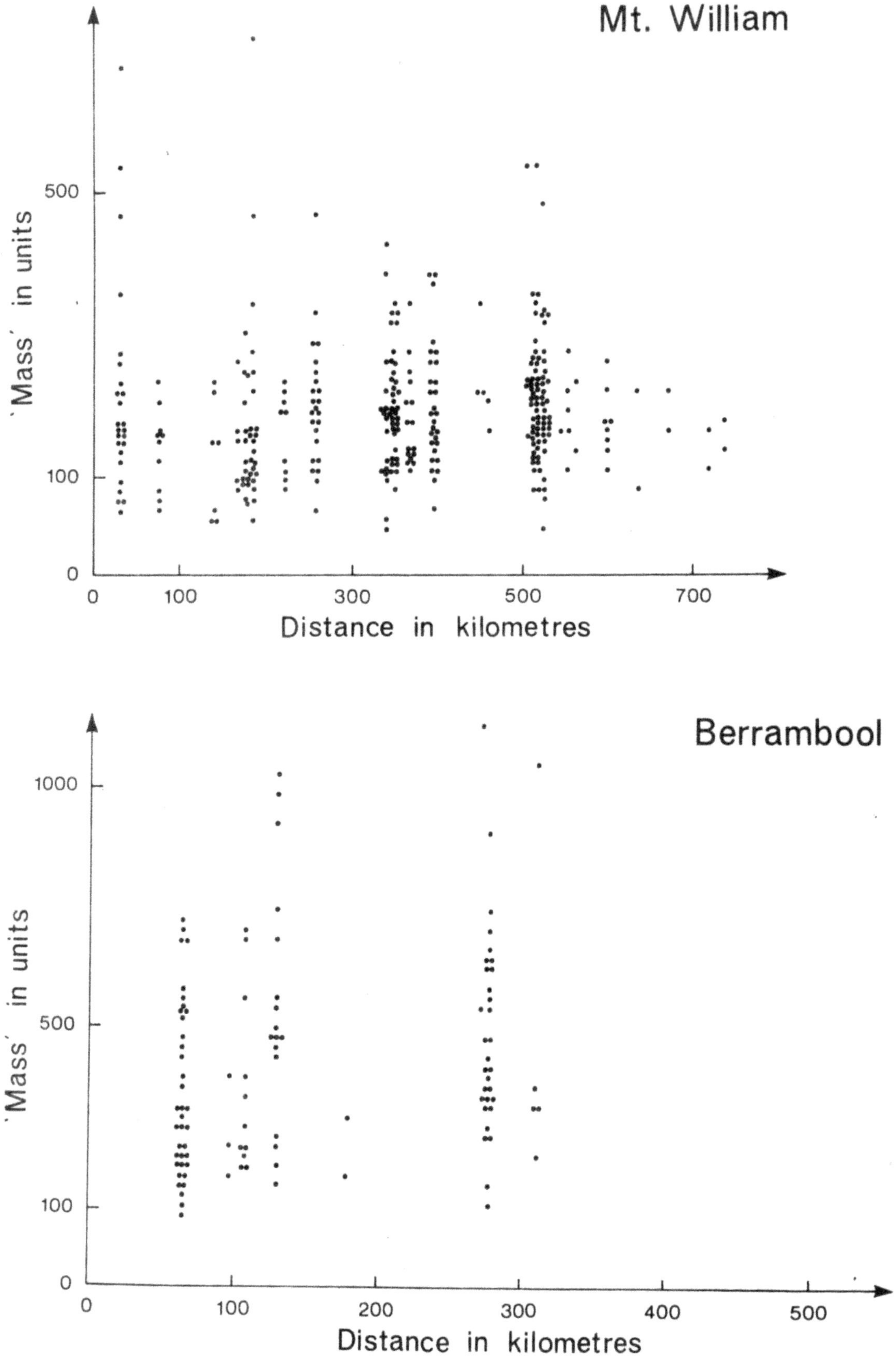


Figure 13.11 Relationship between 'mass' and distance from quarry source for hatchet heads from the Mt William quarry and that at Berrambool.

variations, though slight, could well indicate differences in the perceived value of the stone and differences in the processes of distribution from the Kulin territory to areas beyond this. The increase in size of the Berrambool artefacts over distance seems apparently counter to the hypothesis of early discard until one considers the actual distances involved. All are relatively close to the source, so the trend may not argue against that hypothesis.

These preliminary studies have revealed interesting variations in size, 'mass' distributions that merit more sophisticated examination which will be undertaken in later analyses. In addition, independent approaches will be adopted using relevant but independent artefact attributes in the relative distributions of features such as breakage and re-use, as well as features which reflect the use-history of the artefact, such as heavy damage to working edges, the presence of use-wear, evidence for re-shaping or re-grinding such as skewed working edges. These may add another dimension to studies of the valuable stone and the valued good.

ACKNOWLEDGEMENTS

To the curators of the collections used for this study we owe an obvious debt and we would wish to express our thanks to the curators of the collections in the National Ethnographic Collection, the Australian Museum, the National Museum of Victoria, and the South Australian Museum, as well as of many local collections in Victoria and New South Wales. The project as a whole has been supported by grants from the Australian Research Grants Committee to Isabel McBryde; this is acknowledged with gratitude. Funded by this grant, Alan Watchman worked as research assistant to make the source discriminations essential to the study. The maps and drawings illustrating this paper were prepared for publication by Joan Goodrum of the Australian National University. The photographs were taken by Isabel McBryde.

REFERENCES

Batey, I., n.d. Reminiscences. Manuscript in the library of the Royal Victorian Historical Society, Melbourne.

Bulmer, S., 1964. Prehistoric stone implements from the New Guinea Highlands. Oceania 34(4): 246-268.

Chappell, J., 1966. Stone axe factories in the Highlands of east New Guinea. Proceedings of the Prehistoric Society 32: 96-121.

Coutts, P. J. F. and D. Witter, 1977. New radiocarbon dates for Victorian archaeological sites. Records of the Victorian Archaeological Survey 4: 59-79.

Dawson, J., 1881. The Australian Aborigines. Robertson, Melbourne.

Dickson, F. P., 1978. Australian ground stone hatchets. Ph.D. Dissertation Macquarie University, Sydney.

Groube, L., 1978. Priorities and problems in Dorset archaeology pp. 29-52 In: Darvell, Parker, Pearson, Smith and Thomas (Eds.), New approaches to our past - an archaeological forum. University of Southampton Archaeological Society, Southampton.

Hodder, I. (Ed.), 1978. The spatial organisation of culture. Duckworth, London.

Hodder, I. and Orton, C., 1976. Spatial analysis in archaeology. Cambridge University Press, Cambridge.

Howchin, W., 1934. The stone implements of the Adelaide tribe of Aborigines. Gillingham and Co. Ltd., Adelaide.

Howitt, A. W., 1904. The native tribes of south-east Australia. Macmillan, London.

Hughes, I., 1977. New Guinea Stone Age trade. Terra Australis 3, Australian National University, Canberra.

Lampert, R. J., 1972. Hagen axes. Unpublished typescript. Prehistory Department, Research School of Pacific Studies, Australian National University, Canberra.

Lourandos, H.,1977. Aboriginal spatial organisation and population: south-western Victoria. Archaeology and Physical Anthropology in Oceania XII: 202-225.

McBryde, Isabel, 1978. Wil-im-ee Moor-ring: or, Where do axes come from? Mankind 11: 354-82.

McBryde, Isabel, 1979a. Petrology and prehistory: lithic evidence for exploitation of stone resources and exchange systems in Australia, pp. 113-126 In: Clough, T. and W. Cummins (Eds.), Stone Axe Studies, Council for British Archaeology, London.

McBryde, Isabel, 1979b. Ethnohistory in an Australian context: independent discipline or convenient data quarry? Aboriginal History 3(2): 128-151.

McBryde, Isabel and Watchman, A. L., 1976. The distribution of greenstone axes in south-eastern Australia: a preliminary report. Mankind 10: 163-174.

Renfrew, C., 1977. Alternative models for exchange and spatial distribution. pp. 71-90 In: Earle, T. and Ericson, J. E. (Eds.), Exchange systems in prehistory. Academic Press, New York.

Smyth, R. Brough, 1978. The Aborigines of Victoria, 2 vols. Government Printer, Melbourne.

Strathern, M., 1965. Axe types and quarries. Journal of the Polynesian Society 74(2): 182-191.

Strathern, M., 1969. Stone axes and flake tools: evaluations from New Guinea. Proceedings of the Prehistoric Society 35: 311-329.

Thomson, D. F., 1949. Economic structure and the ceremonial exchange cycle in Arnhem Land. Macmillan, Melbourne.

Watchman, A. L. and Freeman, R. S., 1978. Trace element analysis of Aboriginal greenstone artefacts. Royal Society of Victoria Proceedings 90 (1 and 2): 271-281.

Wright, R., 1979. A modicum of taste: Aboriginal cloaks and rugs. Australian Institute of Aboriginal Studies Newsletter 11: 57-68.

## 14. PATTERNS IN THE PREHISTORIC DISTRIBUTION OF STONE RESOURCES IN THE CHATHAM ISLANDS

Douglas G. Sutton and Hamish J. Campbell

### INTRODUCTION

The Chatham Islands lie 860 km east of the South Island of New Zealand (Figure 14.1). This distance is well beyond the maximum range of two-way canoe voyaging of 600 km proposed by Green (1976: 60). The most likely mode of first settlement of the Chathams is drift voyaging from New Zealand (Levison et al., 1973) but even this is likely to have been possible only before about A.D. 1400 due to the climatic deterioration which occurred after that date(Leach and Leach, 1979: Figure 1; see also Burrows and Greenland, 1979). Therefore, the Polynesians who first settled the islands would have been few in number, and unlikely to receive any significant secondary settlement. They were isolated for at least the 400 years between A.D. 1400 and European discovery of the islands in A.D. 1792. In that interval, and possibly for significantly longer, the archipelago was a closed system.

Moriori culture developed after initial settlement of the Chathams from a New Zealand derived Archaic East Polynesian culture which is represented archaeologically by artefacts recovered from a small number of sheltered east coast positions on Chatham and Pitt Islands. These include all the ornament forms found at the type site of the New Zealand Archaic Culture (Skinner and Phillips, 1953).

The major characteristics of Moriori culture are:

(1) An economic strategy based on coastal hunting of seals and pelagic birds, fishing and the gathering of shellfish. This allowed high population densities to be maintained.

(2) A range of settlement patterns focused on perennial occupation of coastal or lake shore villages and the dispersal of small groups in summer months to littoral and inland resource zones at which specific food resources were taken in quantity. This dispersal did not involve long periods spent away from village settlements. Each village group used resources within a small range of its major settlement.

(3) A material culture characterized by the importance of flake tools, rough subrectangular adzes, bird bone awls, a range of abraders and several forms of patu or handclubs. Fishhooks, bird spears and highly finished adzes are uncommon.

(4) An egalitarian social structure (Sahlins, 1958; Ryan, 1977; Earle, 1978), based on localized kin-based groups ranging in size

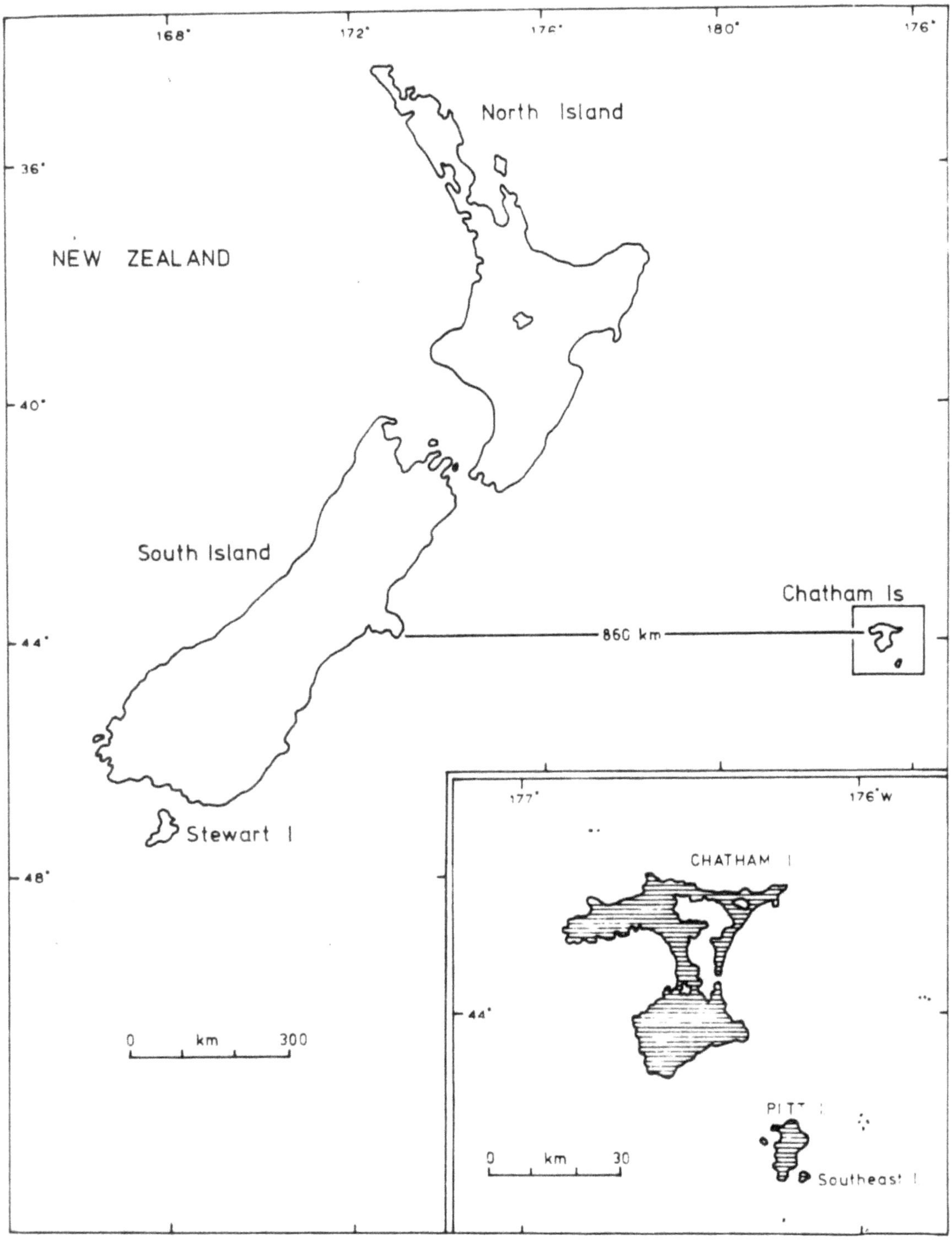


Figure 14.1 Location map: Chatham Islands.

from 30 to 100 individuals. Each of these groups occupied a village settlement within an area to which it had prior, if not exclusive, rights (Deighton, Ms., 1882). "Headmen" (Richards, 1972: 352) were senior members of the kin-based groups. Their arbitrary powers were very limited. Supernatural sanctions were invoked to regulate the exploitation of food resources. Lethal combat was prohibited within and between kin groups.

The view of Moriori culture presented above has been developed by one of the authors (Sutton) on the basis of archaeological and historical evidence, particularly from the reconstruction of subsistence economics and settlement pattern of the 16th century Moriori who occupied the Southwest Coast of Chatham Island. Moriori culture was a notable departure from both the Archaic culture from which it developed, and Classic Maori culture of northern New Zealand. It came into existence prior to A.D. 1500 and was extinct by about A.D. 1860 (Sutton, 1979a).

If the reconstruction offered above is accurate it should be reflected in the redistribution of workable stone materials. Specifically, if Moriori culture was based on localized, self-sufficient groups one would expect the cultural redistribution of stone resources to be confined to within a small distance of their geological sources. Extensive transportation of stone materials, on the other hand, should be represented in Archaic sites on the Chathams. All of these sites identified to date are in sheltered east locations with cove creek mouth harbours.

This paper reports the field and laboratory investigation of workable stone resources in the Chatham Islands. This was undertaken by one of the authors (Campbell). The analysis of excavated and surface collected samples is also reported. Results are interpreted in terms of the problem outlined above.

## THE STONE MATERIALS

The geology of the Chathams can be described in terms of three major zones (Figure 14.2; see Hay et al., 1970 for a full description). These are:

(1) The schist zone which forms the whole of the northern third of Chatham Island. Schist constitutes basement rock underlying all sedimentary and volcanic rocks in the Chathams Rise - Chatham Islands regions. The schist exposed on Chatham Island is thickly mantled by Quaternary peats and sands. In places, minor basalt flows of Cretaceous age and remnants of a thin sedimentary cover of lower to upper Tertiary age are preserved. The low-lying schist-peat landscape is punctuated by two diverging rows of small volcanic plugs and cones of Tertiary age.

(2) The southern volcanic region which includes Pitt Island and the large southern block of Chatham Island. This zone is composed of a thick pile of gently dipping to flat lying Cretaceous olivine basalt, trachy-andesite and minor phonolite flows; breccias, tuffs and plugs of trachyte and basomite. Pitt Island consists of a northward dipping, dissected plateau of Cretaceous volcanics with a single upstanding trachyte dome to the north; a well exposed succession of overlying

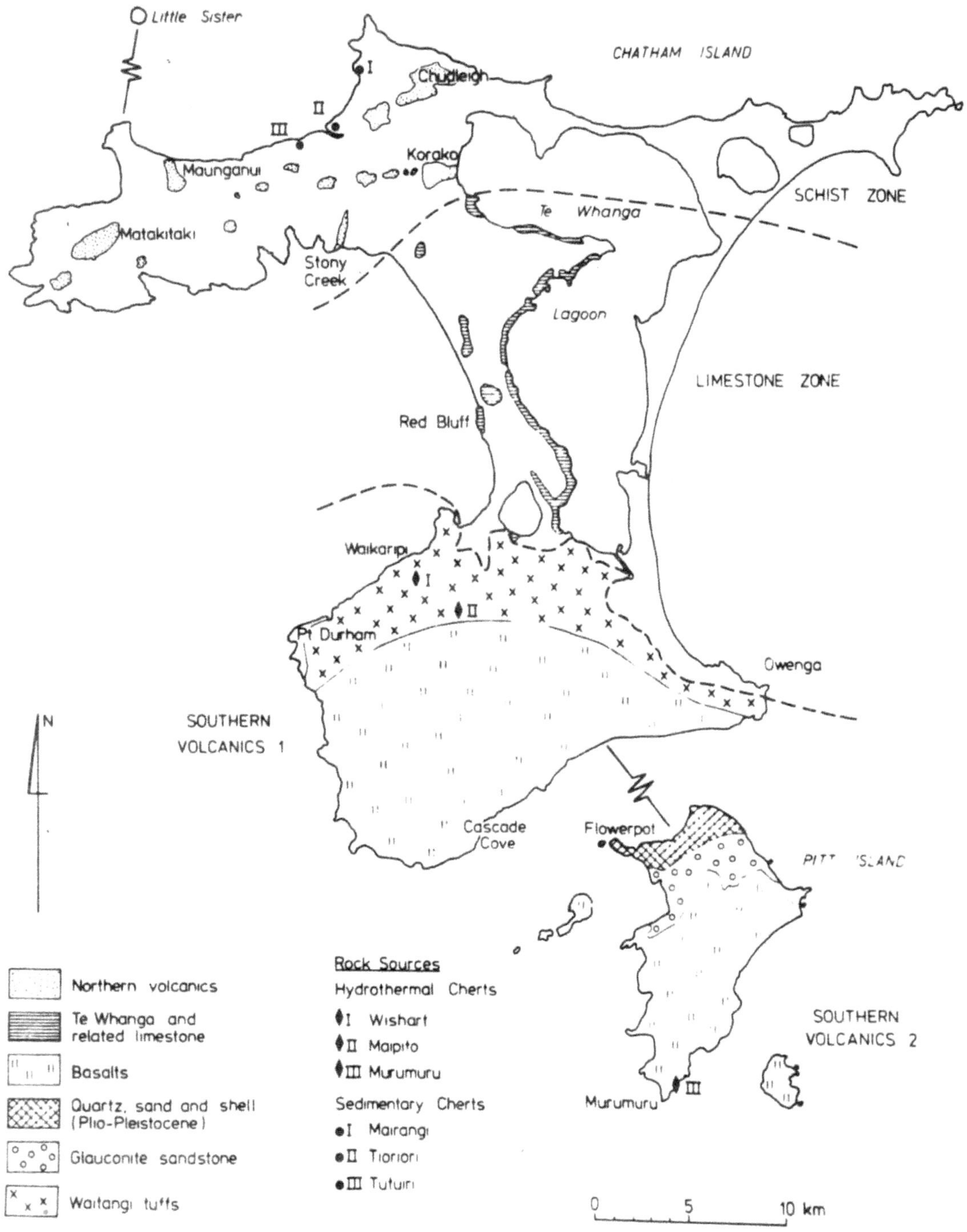


Figure 14.2 Geology of the Chatham Islands.

younger sediments (Upper Cretaceous to Pliocene) to the northwest; and a number of large dissected limburgite breccia pipes outcropping along the west coast and forming many of the offshore islands. The southern block of Chatham Island forms a north-sloping peat-covered landscape entirely underlain by Cretaceous volcanics (flows, breccias and tuffs, minor dykes and sills), which are particularly well exposed along the southern coast.

(3) The middle, sedimentary, zone of Chatham Island includes the area between the southern volcanic zone and the schist zone. The middle zone is composed of tuffs, limestones and sands of Lower Tertiary to Quaternary age which occupy a structural depression between the schist to the north and the Cretaceous volcanics to the south. They are moderately well exposed along the western shore of Te Whanga lagoon and at Red Bluff in Petre Bay. They are comparable to the sediments exposed in the northwest corner of Pitt Island.

The workable stone materials identified (Table 14.1) and characterized by Campbell (n.d.) are distributed throughout these zones (Figure 14.2). The southern volcanic zone contains three hydrothermal cherts; the Wishart and the Maipito sources which outcrop southwest of Waitangi, and the Murumuru material from the southern end of Pitt Island. Pitt Island rhyodacite is from an undiscovered source thought by Campbell (n.d.) to be on the southern end of Pitt Island. The workable dolomitic limestone occurs in a single isolated locality in Cascade Cove. The middle zone contains Te Whanga chert and the poor quality Red Bluff biomicrite.

Materials from within the schist zone include the sedimentary cherts from the related areas of Mairangi, Tutuiri and Tioriori. The workable Takatika grit was found, evidently near its primary source, in the creek mouths of this area. A poor quality hydrothermal chert was found at Korako. A minor chert source was found on the Little Sister, 20 km north of Chatham Island. The tough Wharekauri chert is distributed in small detrital nodules over most of the exposed hard surfaces of the schist zone, particularly in the north and west. It is also found in small quantities at a few locations on Pitt Island. A deposit of very small nodules was found by Campbell (n.d.) near the Waihora site on the southwest coast of Chatham Island.

The materials listed above may be all those available to the Moriori, apart from the basalts and related volcanics. Campbell (n.d.) examined Moriori artefacts held in the Auckland, National, Southland, and Chatham Islands Museums and found no new materials.

Crystalline volcanic stone materials with conchoidal or subconchoidal fracture occur at several places in the islands. These include at least two areas on Pitt Island, the Owenga-Manukau area on the southeast of Chatham Island, several locations on the southwest coast and one area in the northwest. The latter is the Ouira or Stony Creek quarry (N.Z.A.A. Offshore Island Site Number C240/188) at which material from a local columnar sill was worked to form adze roughouts. Hundreds of these can be seen spread out over about half a hectare. It was not possible to source the crystalline volcanic rocks precisely because petrological study of the Chathams basalts has not yet been developed to a point at which source or specific source area characterization

TABLE 14.1 WORKABLE STONE MATERIALS OF THE CHATHAM ISLANDS (after Campbell, n.d.)

(1) Hydrothermal Cherts

Wishart Chert
Maipito Chert
Murumuru Chert
Korako Chert

(2) Sedimentary Cherts and Silicified Sediments

Wharekauri Chert
Te Whanga Chert
Tioriori Chert
Tutuiri Chert
Mairangi Chert
Takatika Grit
Red Bluff Biomicrite

(3) Other Rock Types

Pitt Island Rhyodacite
Cascade Dolomite
Obsidian

is possible. However, rocks from the northern zone, and particularly the fine-grained and distinctive Ouira material, could be separated from rocks from the southern volcanics. This distinction proved to be very useful and further work is in progress which may allow more specific identification of source.

Obsidian flakes have been recovered from several areas of the Chatham Islands, but no local source has been found despite a sustained search. Over 70 worked pieces were found by the author and others between 1974 and 1976. Sixty of these were found on two Pitt Island sites which appear to be Archaic. A long chisel in argillite from a Nelson mineral belt source in New Zealand was also found in one of these sites. The obsidian is now being analysed at Lucas Heights by Dr Roger Bird. It is excluded from the present discussion because results of this work are not yet available.

## EVIDENCE OF REDISTRIBUTION

### The Durham Evidence

Detailed evidence of patterns in the cultural redistribution of the 15 stone materials described above is available from excavated sites in the Durham area, southwest coast of Chatham Island (see Figure 14.3 for place names). One central place village site (known as the Waihora mound site), four specialized coastal middens and two inland fowling sites were excavated. They are functionally inter-related parts of a single contemporary settlement pattern. Worked stone was recovered from five of the six sites. The number of flakes assigned per source from each site is shown in Table 14.2, with the minimum distance from Waihora to each source. Flakes included in the study were both debitage chips and larger used flakes. More fully prepared non-basaltic artefacts were excluded. In view of the limitations of the petrology of Chathams basalts no attempt was made to source the excavated basalt flakes. Instead, the 20 finished basalt artefacts found in Waihora were thin-sectioned and the sections later examined by Paul Morris, Geology Department, Victoria University, Wellington. He identified 13 of the artefacts as being from southern volcanics; the other seven are from the Ouira source.

The inland and coastal middens contained few flakes because they represent short-term specialized exploitation of a specific resource rather than sustained occupation (Sutton, 1979b). The village settlement, on the other hand, contained a very large assemblage of flakes (n=10, 029). They were recovered from within the single cultural level at the site which represents occupation during about a century around A.D. 1550. The flakes were distributed in a non-random pattern between the various activity areas which comprised the village (Sutton, 1979a).

The lithic assemblage from Waihora (Table 14.2) is highly structured. Only a proportion of the potential sources are represented. Some sources are much more frequently represented than others. Three factors can be identified which explain the pattern in this variation. The first is distance from the source. The second is the quality of the material in question. The third factor, which is obviously not independent of the second, is the purpose for which a specific material was used. Each of these factors is discussed briefly below. The overall pattern they form is described at the end of this section of the paper.

TABLE 14.2 MINIMUM DISTANCE TO SOURCES AND RELATIVE FREQUENCY OF THE STONE TYPES IN THE FLAKE ASSEMBLAGE FROM THE WAIHORA SITE

| Material | Minimum Distance (km) | Number | Percentage |
|---|---|---|---|
| 1. Basalt | Unknown | 5363 | 53.47 |
| 2. Wishart Chert | 8.8 | 3145 | 31.36 |
| 3. Maipito Chert | 10.4 | 0 | |
| 4. Cascade Dolomite | 13.6 | 301 | 3.00 |
| 5. Limestone | 15.2 | 16 | 0.16 |
| 6. Red Bluff Biomicrite | 17.6 | 3 | 0.03 |
| 7. Te Whanga Chert | 27.2 | 0 | |
| 8. Korako Chert | 28.8 | 0 | |
| 9. Takatika Grit | 29.6 | 0 | |
| 10. Tioriori Chert | 29.6 | 46 | 0.46 |
| 11. Tutuiri Chert | 30.4 | 0 | |
| 12. Wharekauri Chert | 32.1 | 873 | 8.70 |
| 13. Mairangi Chert | 34.4 | 0 | |
| Materials of Unknown Sources: | | | |
| Schist (could be local) | | 272 | |
| Pumice (waterborne) | | 3 | |
| Carbonaceous Residue | | 4 | 2.72 |
| Calcite Crystal | | 3 | |
| Quartz Sand | | 1 | |
| | Totals | 10,030 | 100.00 |

The distance factor is a general background effect. In the non-basalt flake material there is an inverse correlation between distance from source and frequency of occurrence (Table 14.2). The Wishart is the closest and most frequently represented non-basalt. The northern chert sources tend to be represented only very infrequently.

Consideration of the quality of the material seems to account for some of the anomolies in the distance/frequency correlation. It explains the absence of a group of materials in the middle distance range. They are the non-dolomitic limestone, Red Bluff biomicrite, Te Whanga chert and the Korako chert. They are all soft, splintery or brittle (Campbell, n.d.). Similarly the Maipito chert which outcrops just beyond the Wishart source is absent from the Waihora site, possibly because of its poor working quality (Campbell, n.d.). On the positive side, the exceptional hardness and edge-holding ability of the Wharekauri chert appears to account for its relatively common occurrence in the site while all the other northern materials (numbered 9-11, 13 in Table 14.2) are very uncommon.

The function of tools made from a specific material was also important. Some materials were transported from distant parts of the islands to be used for specific purposes. This appears to have occurred because the distinctive physical properties of a particular material were necessary to a specific task. A clear example of this is reflected in the source data on the basalt artefacts from Waihora. As Campbell (n.d.) has noted, the adzes of southern stone types were made of a local trachyte and,

> "The trachyte lithology of the artefacts is identical to the trachyte dykes which occur between Point Durham and Point Weeding Waitangi. Trachyte dykes are not known to occur elsewhere in the Chatham Islands. The trachyte....is not a particularly workable stone."

The seven Ouira basalt adzes may therefore have been imported to overcome deficiencies in the local stone. They include the finest and sharpest adzes in the site (Sutton, 1979a). The common occurrence of Wharekauri chert in Waihora, already mentioned above, is a second example of this sort. Others are found in the materials used in the abraders. These include sandstone and shale, neither of which is available in the vicinity of Waihora. These artefacts are illustrated and described in Sutton (1979a).

It appears that some stone types tended to be used for specific functions. For instance, at least two schist handclubs are known from the Waihora site. Another example was excavated by one of the authors (Sutton) from a nearby and contemporary site. The stone used in these could only have come from the schist zone almost 30 km to the north. As another example, a total of six adzes and four "chisels" (Skinner, 1923) of dolomitic limestone were found in the site, although this distinctive material comprised only three percent of the flake assemblage. The hard limestone was, therefore, favoured for use in adzes and not in flake tools. Most of these were evidently made of the Wishart chert. Several mataa (Skinner, 1923) and many other flake tools in this material are known from the site (see also Table 14.2).

The overall pattern which emerges is one in which quite large quantities of stone materials which were available within a short distance of the Waihora site were used there, most frequently as flake tools. These materials were the local chert and various crystalline volcanic rock types. They are of good to intermediate quality. Small quantities of preferred stone materials were imported from distant sources and used in functions for which the local materials were not well suited. Most of the lithic needs of the inhabitants of Waihora appear to have been met by the flake tools in chert. Adzes and particularly more fully finished examples are uncommon.

## The Broader View

There are two pattern of redistribution evident in the Chathams. The first is associated with the few putative Archaic sites known in the islands. They have been identified in the field or the literature by the presence of Archaic ornament forms and large rectangular and hogback adzes. The second pattern is associated with non-Archaic sites in all areas of the islands. These are identified in the field either by the presence of artefact forms which are characteristic of Moriori culture (Skinner, 1923) or by the small and highly specialized nature of the many seasonal, short term sites present in Moriori settlement patterns (Sutton, 1979a). The Archaic pattern comprises evidence of extensive redistribution involving materials from New Zealand, and possibly elsewhere, and widely separated areas of the Chathams. The later pattern is of localized redistribution of stone resources. This is essentially the pattern reconstructed for the Waihora site.

The Archaic pattern is best represented by the materials found in two Archaic sites on Pitt Island; the Tupuangi river-mouth site (C240/619) and the Waipaua Rivermouth site (C240/570). There are several large worked boulders of Takatika grit on the site at Tupuangi. They were evidently taken there by canoe rather than the wash-through raft associated with Moriori culture (Skinner, 1919). These are known to have had limited weight-bearing and seaworthiness properties (see Jones, 1976) and were used in short drift voyages and not long voyages against current and wind represented by the Takatika grit at Tupuangi. Sixty pieces of obsidians and a long chisel in D'Urville Island argillite have also been found in the Archaic sites on Pitt Island. Both come from beyond the Chathams. The source of the obsidian is not yet established (see Leach and Warren this volume). There are also some Cascade dolomite adzes and flakes of Wishart chert present. They reflect voyages across Pitt Strait in the early period of Chatham Islands prehistory.

The Moriori pattern, on the other hand, is exemplified in the localized redistribution of four different stone materials which is apparent in evidence surface collected from non-Archaic sites. The materials are: dolomite limestone; Takatika grit; Tioriori chert and Pitt Island rhyodacite.

The Cascade limestone is known to have been found in the Waihora site and only two other locations in the Chatham Islands. Smith (1892) reported an adze in the limestone from Owenga. Campbell (n.d.) has collected a few flakes from the Red Bluffs. No non-Archaic occurrences are known from south of the Pitt Strait.

Flakes of Takatika grit are common in surface collections from the many creek mouth and sand dune middens in the northwest of Chatham Island. The only occurrence of the material outside this small area apart from the boulders in Archaic contexts on Pitt Island, are at Awapatiki and Red Bluff (Campbell, n.d.).

Tioriori chert occurs naturally in large blocks in the northwest of Chatham Island between the Tioriori and Takatika Rivers. These blocks were worked and the secondary distribution of the material is concentrated in the local area. The material is prevalent (Campbell, n.d.) in surface collections from sites between Cape Young and Maunganui. There are only three known occurrences outside this small area. The most distant of these is the few flakes found in the Waihora site (Table 14.2). The other two were small flakes found by Campbell (n.d.) at Ocean Bay and Whangatete Inlet about 4 km south of the source.

Finally, Pitt Island rhyodacite has been surface collected from several sites on the southern half of Pitt Island and on nearby Southeast Island. It is the only worked stone material found on Round Island in the Star Keys group, about 13 km east of Pitt Island. It is represented north of the Pitt Strait by only three flakes which were found together at the mouth of the Tuku River in the extreme southwest of Chatham Island (Campbell, n.d.; Figure 14.3). This evidence of the very limited non-Archaic redistribution of these four materials concludes the data section of the paper.

## DISCUSSION AND CONCLUSION

Evidence for the division of the Chatham Islands into "tribal areas" (Richards, 1962, 1972; Simmons, 1962, 1964) at least during the latter or Moriori phase of the prehistoric period can be seen in both the archaeological evidence and historical records. These areas coincide with the limited non-Archaic distributions of stone materials. This suggests that socio-political boundaries, as well as physical considerations, set the limits of redistribution.

Between 1974 and 1976 over 700 archaeological sites were recorded in the Chatham Islands. During research for this paper the site records were searched for evidence associated with major settlement sites but not found in the specialized middens. This included site size over about 300 square metres, evidence of houses and other discrete activity areas, human skeletal remains, a range of stone types and portable artefacts and man-made pits.

Two or more of these criteria occurred together in only 17 locations (Figure 14.4). The presence of at least this number of major prehistoric settlements is therefore indicated. The field survey evidence has some obvious limitations. First, the identified sites represent an unknown proportion of those present. Secondly, without excavation there is no way of dating each of the sites satisfactorily. It has, however, been possible to exclude the Archaic sites mentioned earlier in the paper and the historic Maori and Moriori villages listed by Simmons (1964). The major settlement sites identified are therefore taken, at least tentatively, to date to within the Moriori phase of Chatham Islands prehistory. If they were, in fact, contemporary we may expect a maximum of 100 people to have resided at each centre. The distribution of these major sites is of considerable interest. They are closest together on the north coast and along the western shore of the very productive Te Whanga Lagoon and are most commonly found in ecotonal situations such as those in which rocky and sandy shores juxtapose. There is also a high correlation between the locations of these sites and former or extant fur seal colonies (Sutton, 1979a).

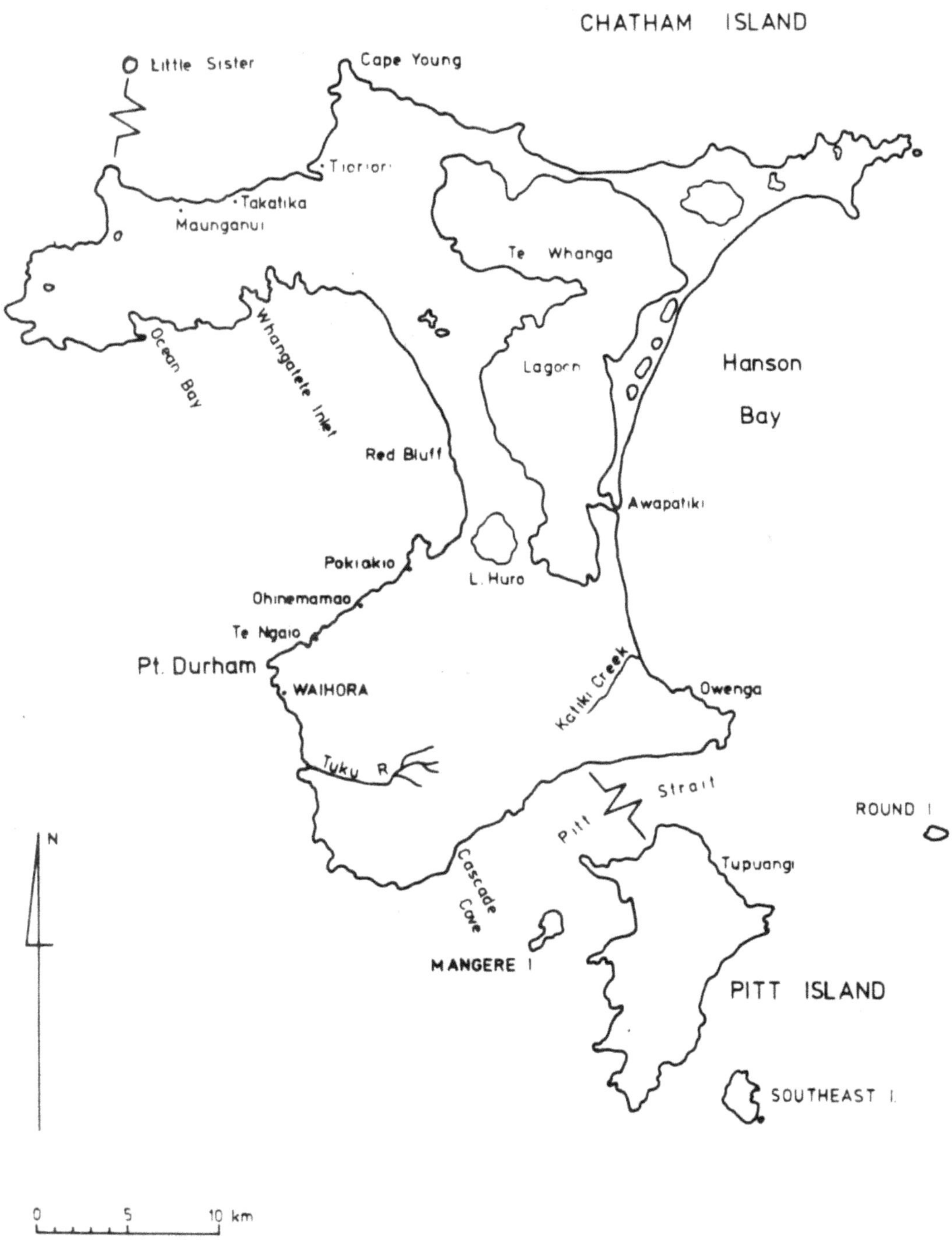


Figure 14.3 Place names in the Chatham Islands.

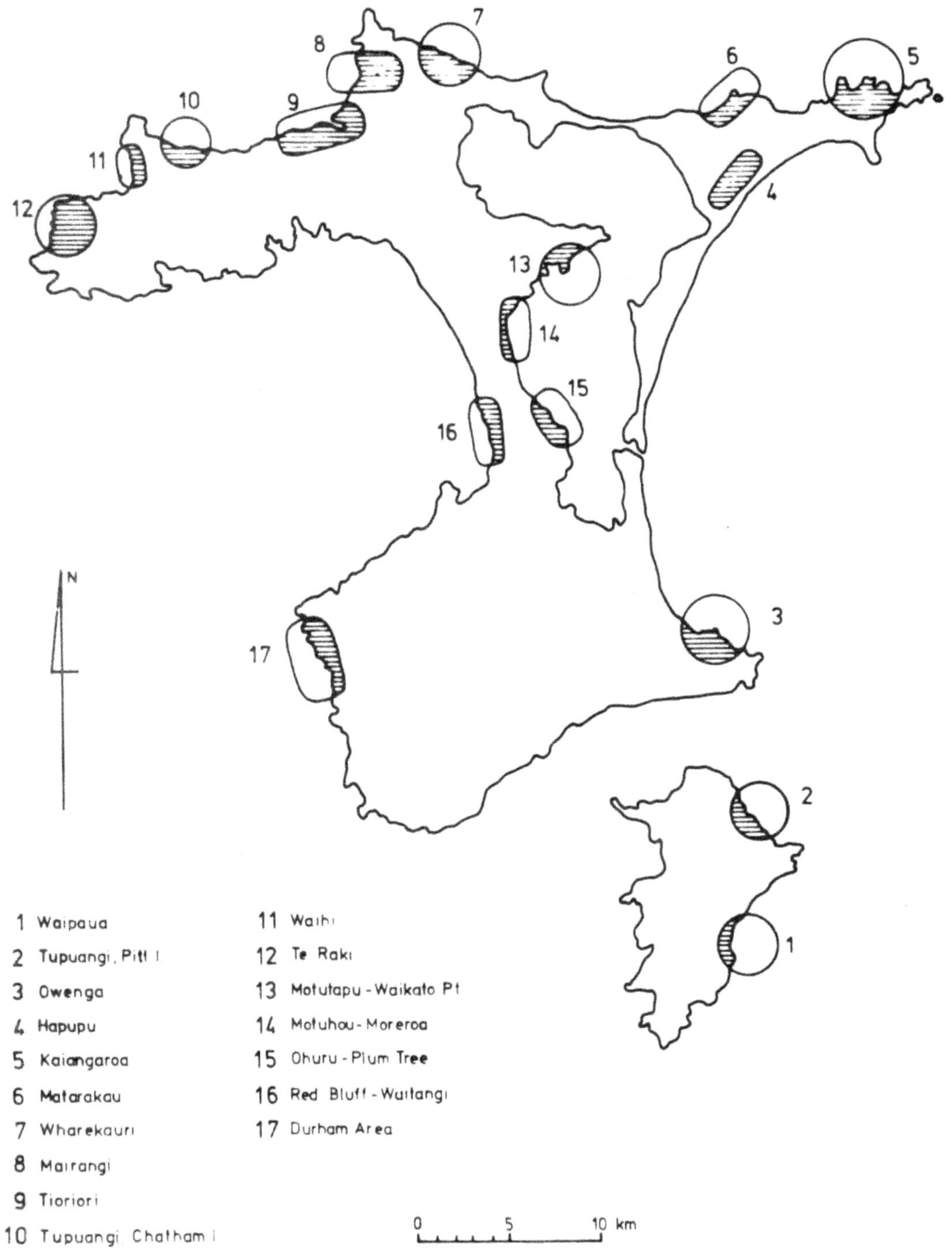


Figure 14.4 Major prehistoric settlement sites in the Chatham Islands.

The purpose of this paper has been to use data on the cultural redistribution of stone materials to assess the possibility that the Moriori lived in essentially self-sufficient, localized groups. The recurrent pattern of limited redistribution, specialization in the use of locally available materials and the very limited volume of trade or exchange occurring between separate areas during the latter part of the prehistoric sequence, confirms this possibility. It is further supported by archaeological evidence for the presence of a small number of major settlement sites and the division of the islands into "tribal areas".

ACKNOWLEDGEMENTS

The authors wish to thank Paul Morris for examining the thin sections and Graeme Mason, Anthropology Department, University of Otago for the use of thin sectioning equipment.

REFERENCES

Burrows, C. J. and Greenland, D. E., 1979. An analysis of the evidence for climatic change in New Zealand in the last thousand years. Evidence from diverse natural phenomena and from instrumental records. Journal of Royal Society of New Zealand 9(3): 321-373.

Campbell, H. J., n.d. The natural and cultural distributions of workable stone materials in the Chatham Islands. Manuscript, Geological Survey, D.S.I.R., Lower Hutt.

Deighton, S. J., 1882. Manuscript in the John White Papers, B75/27. Alexander Turnbull Library, Wellington.

Earle, T., 1978. Economic and social organisation of a complex chiefdom: the Halelea District, Kaua'i, Hawaii. Anthropological Papers, Museum of Anthropology, University of Michigan 63. Ann Arbor, Michigan, University of Michigan.

Green, R. C., 1976. New sites with Lapita pottery and their implications for an understanding of the settlement of the Western Pacific. pp. 55-87 In: La préhistoire Océanienne, Prétirage, Colloque XXII, U.I.S.P.P. IXe Congrès. C.N.R.S., Paris.

Hay, R. F., Mutch, A. R. and Walters, W. A., 1970. Geology of the Chatham Islands. New Zealand Geological Survey Bulletin n.s. 83.

Jones, R., 1976. Tasmania: aquatic machines and offshore islands. pp. 235-264 In: Sieveking, G.de G., I. H. Longworth and K. E. Wilson (Eds.), Problems in social and economic archaeology. Duckworth, London.

Leach, H. M. and Leach, B. F., 1979. Environmental change in Palliser Bay. pp. 229-240 In: Leach, B. F. and H. M. Leach (Eds.), Prehistoric man in Palliser Bay. National Museum of New Zealand Bulletin 21.

Levison, M., Ward, R. G. and Webb, J. W., 1973. The settlement of Polynesia: a computer simulation. Australian National University Press, Canberra.

Richards, R. M., 1962. An historical geography of the Chatham Islands. M.A. Thesis, Geography, University of Canterbury.

Richards, R. M., 1972. A tentative population distribution map of the Morioris of Chatham Island, circa 1790. Journal of the Polynesian Society 81(3): 350-374.

Ryan, T. F., 1977. Prehistoric Niue: an egalitarian Polynesian society. M.A. thesis, Anthropology, University of Auckland.

Sahlins, M. D., 1958. Social stratification in Polynesia. American Ethnological Society Monograph 29. University of Washington Press, Seattle.

Simmons, D. R., 1962. The Moriori of the Chatham Islands. New Zealand Archaeological Association Newsletter 5(1): 238-244.

Simmons, D. R., 1964. Chatham Island archaeological survey. New Zealand Archaeological Association Newsletter 7(2): 51-59.

Skinner, H. D., 1919. Moriori sea-going craft. Man 19: 65-68.

Skinner, H. D., 1923. The Morioris of the Chatham Islands. Memoirs of B.P. Bishop Museum 9(1). B.P. Bishop Museum Press, Honolulu.

Skinner, H. D. and Phillipps, W. J., 1953. Necklaces, pendants and amulets from the Chatham Islands and New Zealand. Journal of the Polynesian Society 62: 169-195.

Smith, S. P., 1892. Stone implements from the Chatham Islands. Journal of the Polynesian Society 1: 80-82.

Sutton, D. G., 1979a. Polynesian coastal hunters in the subantarctic zone. Ph.d. thesis, Anthropology, University of Otago.

Sutton, D. G., 1979b. Inland and coastal fowling strategies of the prehistoric Moriori . pp. 122-139 In: Anderson, A. J. (Ed.), Birds of A Feather, British Archaeological Reports, International Series, No. 62.

# 15. ARCHAEOLOGICAL ARTEFACT DISPERSALS: A NEW WAY TO TEST THEM FOR EVIDENCE OF PREHISTORIC TRADE

Tigger Wise

In this paper a new approach to the analysis of prehistoric trade is described. Usually, when prehistoric trade is suspected (that is, when the artefacts in question look 'foreign' to their context), archaeologists concentrate on analysing the spatial aspects of the artefact dispersal. Firstly, they record and plot the find-spots, or provenances, of the artefacts they have found scattered over a wide area, and try to reconstruct trade directions. Secondly, they take samples of the raw materials used to manufacture the artefacts, and match these with known sources of those raw materials, such as stone quarries, again to try to reconstruct trade directions. In this way, they hope to establish a firm hypothesis that trade has occurred.

I believe that, although sourcing and distribution studies are essential prerequisites to any trade analysis, they leave out one critical aspect of trade. Trade is not merely the movement of goods. It is the exchange of goods between human groups. Therefore, evidence of transaction is needed also—evidence that goods have actually changed hands—before we can have a strong hypothesis of trade. Otherwise, as Adams (1974) points out, trade and diffusion become almost equivalent terms in the archaeological literature. If we limit our enquries simply to the movement of goods when studying prehistoric trade, we impoverish our analysis, and reduce it to an examination of what went where; and worse, we risk making an outright error. All movement of goods is not trade. Sometimes, movements of goods are due to diffusion, such as when a group of people, with their belongings, moves across the country. Sometimes, people obtain raw materials from a distant spot by free access to the source, and bring them home to manufacture their own artefacts. Neither of these situations can be called trade.

In this paper, I present a test which may help, in some cases, to distinguish between the simple movement of artefacts, and the trade of artefacts. The test is designed to detect whether transaction took place: that is, whether the finished goods passed from the human group that made them, to the human group that used them. It is an extra test, to be added to the spatial tests, not to replace them. In fact, this test needs sourcing and distribution studies before it can be applied. Let me describe the idea on which the test is based.

In Figure 15.1, we see free access, or non-trade. In the top portion of Figure 15.1, the people from area A, area B and area C go to get raw materials from a widely dispersed source. Perhaps they are getting stone from the local rivers to make into axes. Perhaps they are getting amber to bring home to make into necklaces. In the bottom portion of Figure 15.1, we see free access also, but this time, free access to a more specific source. Perhaps the people from area A, area B and area C are sailing to an uninhabited island

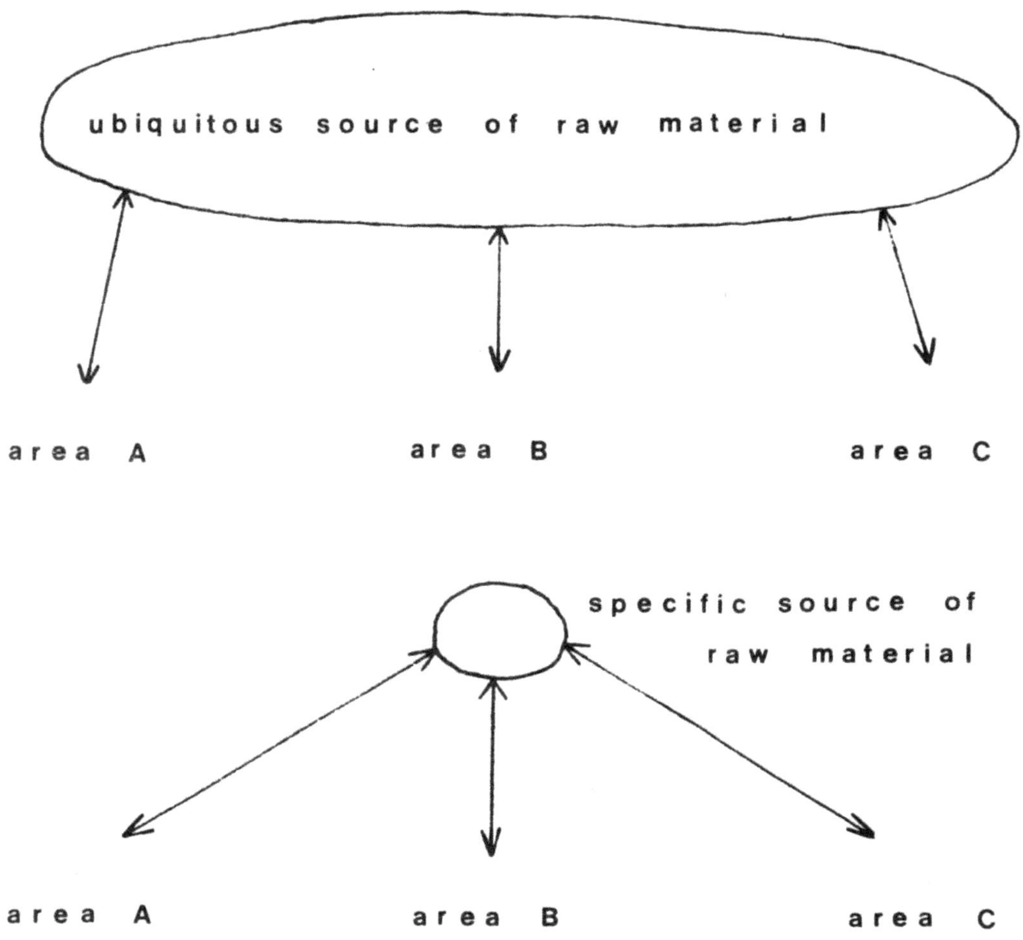


Figure 15.1 Free access exploitation of raw material.

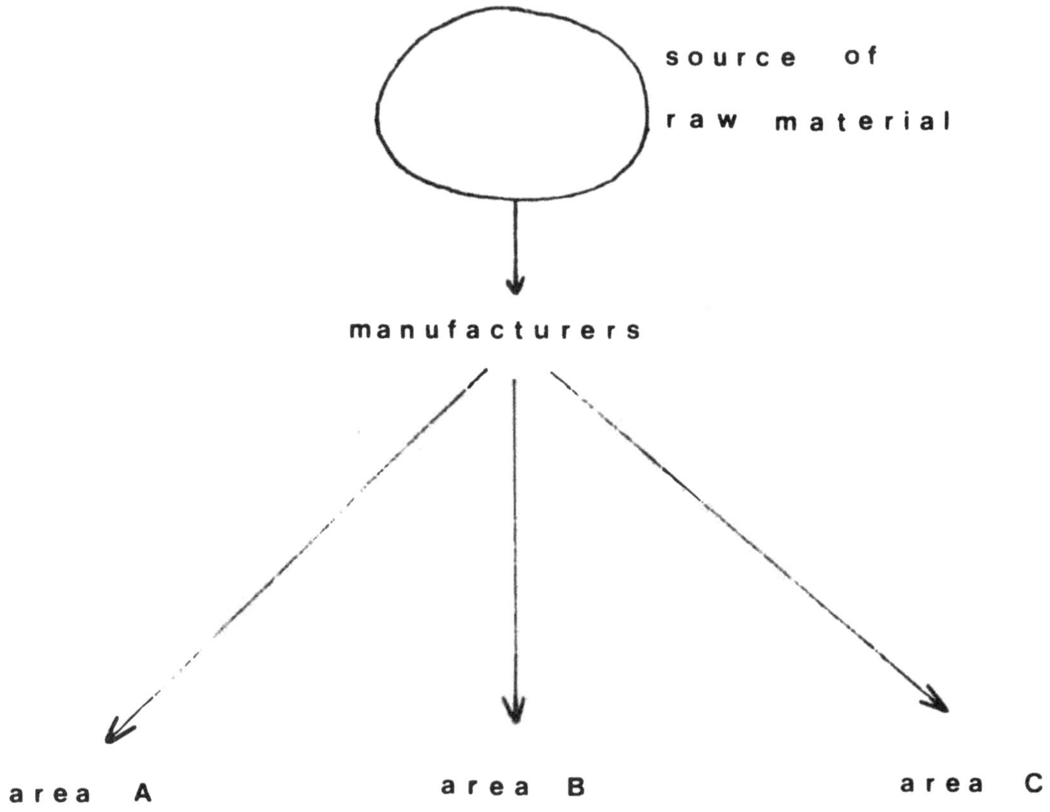


Figure 15.2 Trade in finished artefacts.

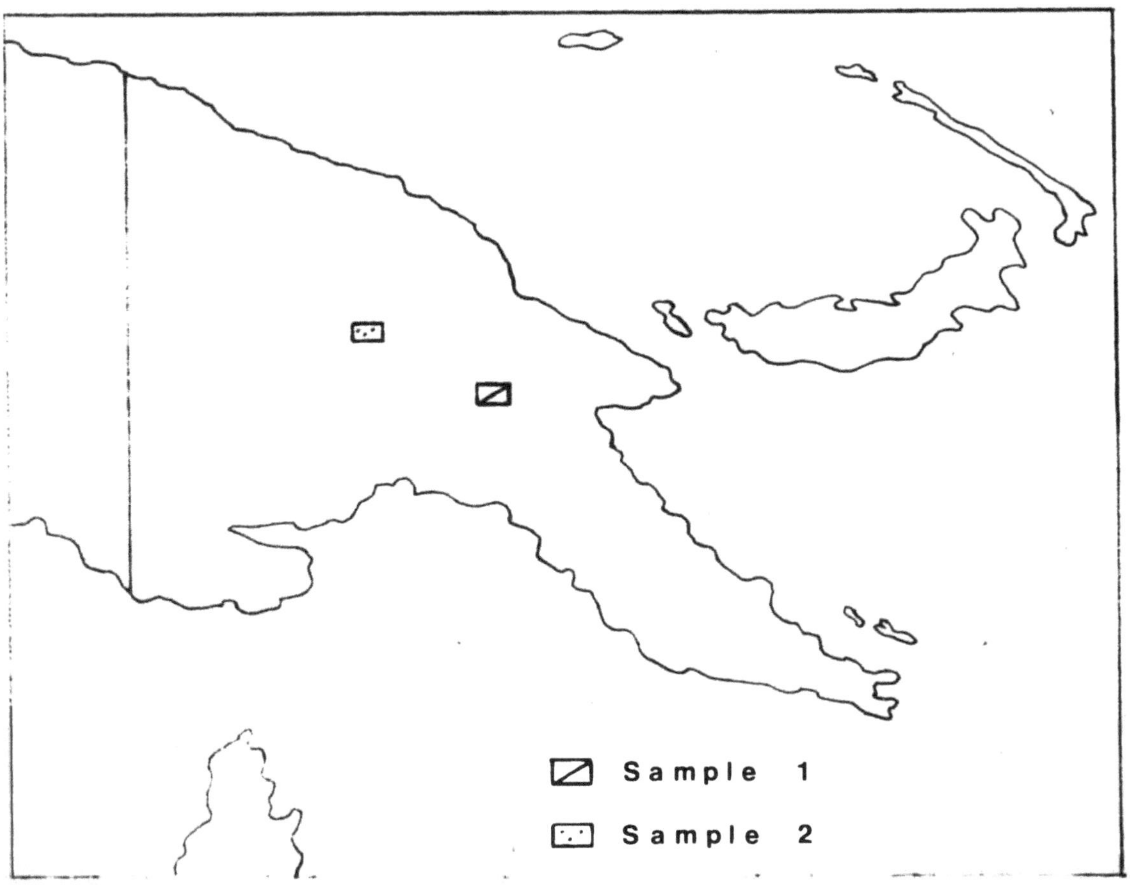


Figure 15.3 Map of Papua New Guinea showing the areas from which the two axe samples were collected.

to get obsidian to bring home to make into blades (see Renfrew, 1972). Perhaps they are going to a greenstone quarry deep in the mountains to get greenstone to bring home to make axes. In both sections of the diagram, the consumers, that is the users, of the artefacts are obtaining their own raw materials, and making their own artefacts. This is not trade. Yet if we confined our analysis to sourcing and distribution studies alone, we might propose the hypothesis of trade.

In Figure 15.2 we do have trade. Owners, or entrepreneurs, have taken over the source of the raw material—the stone quarry, or obsidian mine, or amber bed—and they are extracting the raw material, making the artefacts, and trading the finished goods to area A and area B and area C. This time, after sourcing and distribution studies alone, the hypothesis of trade would be correct. These artefacts have been traded.

So although the spatial distribution of the artefacts is the same in these two cases; the two situations are very different. In the first, we have free access; in the second, we have trade. The problem is: how to tell the difference in the archaeological record? The solution is: examine the parts of the process where the differences occur. The differences occur not in the parts of the process concerned with extraction of raw materials, nor with the final resting place of the artefacts. The differences occur in the production process. Consequently, we must use this production variable to help diagnose trade. Look at the two situations again, this time concentrating on the difference in production between the two.

In Figure 15.1, the manufacturing, or production, was being done by three different groups of people; although all were using the same raw materials. One could predict, then, that if one examined all the artefacts found at area A, area B and area C, those artefacts would fall into three stylistic groups, corresponding to the manufacturing ideas and techniques associated with each of the three areas. The axes from area A would form one stylistic group, the axes from area B would form another stylistic group, and the axes from area C a third stylistic group.

In Figure 15.2, the artefacts were all extracted and made by the same group of people. This time, although the artefacts are scattered over the three different areas, i.e. A, B and C, one could predict that they would be stylistically similar. One could predict that they would not form three groups corresponding to provenance. They would be stylistically homogeneous.

The test I propose in this paper will help to distinguish between these two situations in the archaeological record. The test will produce evidence to show whether the consumer of the goods was or was not the maker of the goods, thus allowing us to hypothesize a little more confidently whether trade has taken place. Obviously, the models show a simplification of the situation. There would be far more variations and complications than the simple free-access versus trading situations that I have described. However, this is the very beginning of this line of investigation.

The test proposed is simple: first, obtain the collection of artefacts you think has been traded. Second, have the usual sourcing tests done on the artefacts. Then select for the test those artefacts which derive from the same source of raw material. Collect a sample of artefacts from each of three, separate areas: a sample found in area A, a sample from area B, a sample from area C. Combine them into one large group. Compare all artefacts, each with each, for stylistic similarities, using multivariate analysis. Arrange them in a sorted similarity matrix. Read the results as follows: if the artefacts cluster into groups that correspond with their find-spot, the artefacts were probably not traded, but made locally, by the local consumers. If the artefacts do not form three stylistic groups corresponding to provenance, then possibly the axes were made by a single group of people, possibly the quarry owners, and traded into the three areas. It is a simple test, easily performed by anyone with access to a computer. The method is dependent on the computer because of the large number of stylistic variables that must be used to compare the artefacts. Once refined, it will be a test that can be easily added to the repertoire of tests for trade.

It is too early to use the test on unknown material in order to ascertain whether or not that material was traded. First, the test itself must be tested under controlled conditions when the answers are known in advance in order to see whether it can discriminate between the two types of situation. Consequently, I am in the process of giving the test a number of trials on known data. It is hoped that this will reveal what types of data it works best on, and how best to interpret the results.

As an initial trial run, I used some stone axe data from New Guinea. These were chosen for a number of reasons: firstly because they were artefacts which had been ethnographically recorded as having been dispersed over the countryside in the two ways I wished to contrast: namely Central Highlands axes have been observed by nearly all anthropologists who have worked in the region (Bulmer, 1964; Strathern, 1966; 1969; Hughes, 1977) to have been traded; whereas Eastern Highland axes have been reported as having been made by each man for his own use, from local river boulders (Blackwood, 1950; Chappell, 1966; Bulmer, 1975). In other words, they were axes obtained by free-access, not trade. Secondly, I selected axes from New Guinea for the trial test because it has been ethnographically observed in the New Guinea Highlands, both Central and Eastern, that the stone axe is an all-purpose tool (Hughes, 1977; Bulmer, 1964; Blackwood, 1950). This means that variations of style are unlikely to be due to variations in the function of the axe; but rather, will reflect varying manufacturing styles. Thirdly, I chose these axes because they had already been sourced. Chappell (in Lampert, 1972) had sourced the Central Highlands group to Abiamp quarry, whereas the Eastern highlands group were of local river boulders. Fourthly all the axes were from surface collections, which gave some measure of control over the time factor.

## DESCRIPTION OF THE TRIAL TEST

A sample of axes of the traded variety had to be compared with a sample of axes of the non-traded variety. Both samples had to have a wide enough dispersal for three clear provenances in each sample to emerge.

Sample One (the example of non-traded axe dispersal) consisted of 50 axes from the Eastern Highlands (see Figure 15.3). The axes in this group were made locally from local stone, by the local users of the axes. The axes were collected from three separate provenances within the Eastern Highlands,

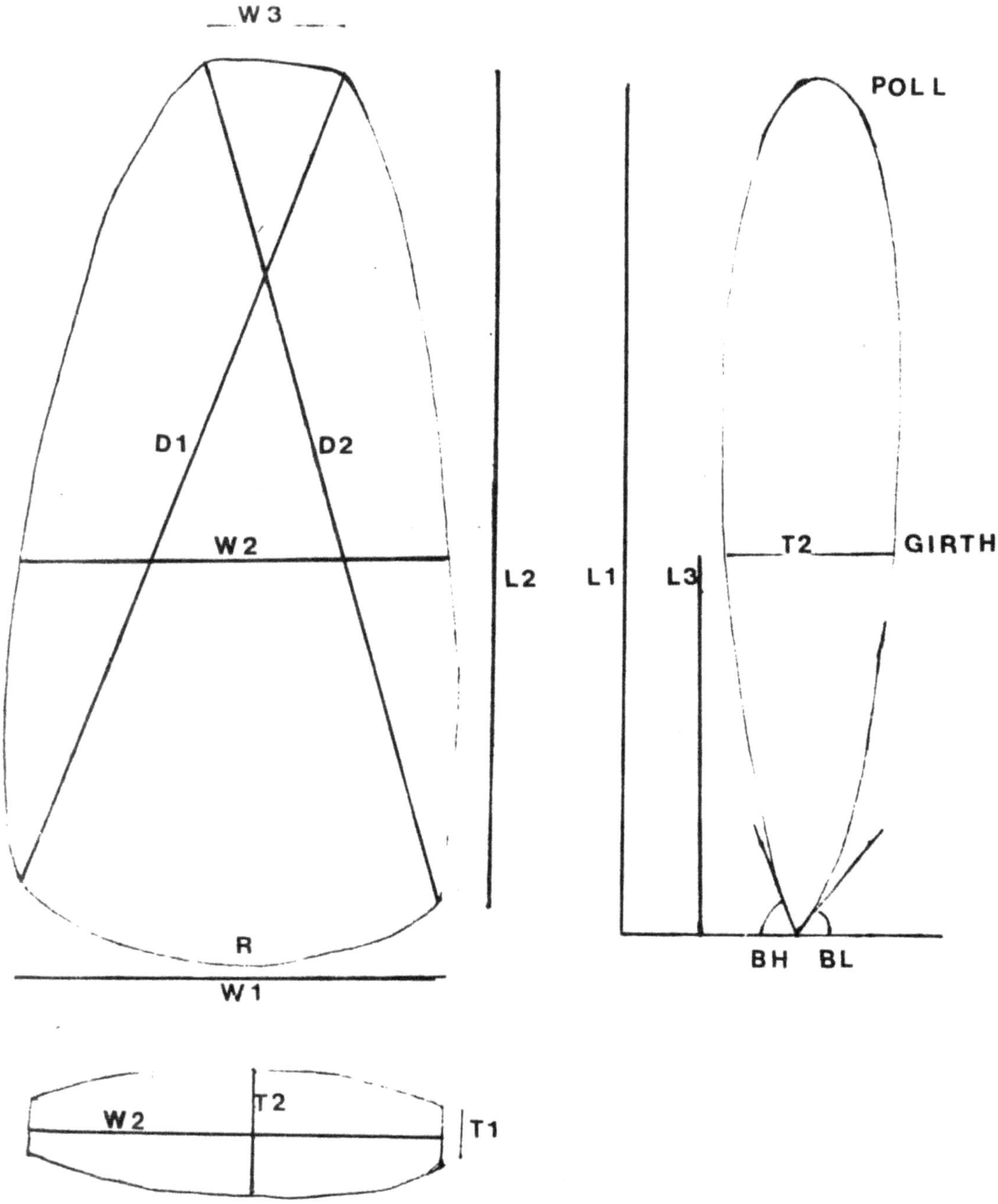


Figure 15.4 Diagram of an axe showing measurements taken, and the terminology used in the text (after Lampert, 1972).

by a number of different collectors since 1966. They are housed at four locations: The University of Papua New Guinea: J. J. McCarthy Museum, Goroka; Papua New Guinea Art Gallery and Museum, Port Moresby; and the Australian Museum, Sydney. I used 17 axes from the Arona area, 21 axes from the Wonenara area and 12 axes from the Okapa area. These three areas in the Eastern Highlands are over 45 km apart, and are separated from each other by deeply ravined, rough mountain country with few roads.

Sample Two (the example of traded axe dispersal) consisted of 72 axes from the Central Highlands (see Figure 15.3). The axes in this group were all sourced to a stone quarry called Abiamp in the Central Highlands, from which they were widely traded in the area (Chappell, 1966; Chappell in Lampert, 1972). The axes were collected from three separate provenances within the Central Highlands by Lampert (1972). They are housed in the Papua New Guinea Art Gallery and Museum, Port Moresby. I used 33 axes from the Wahgi area, 32 axes from the Nebilyer area, and 7 axes from the Kaugel area. These three areas in the Central Highlands are over 40 km apart, again separated by deeply ravined, rough mountainous country. The three 'areas' of each of these two samples are simply 'areas': no information about their cultural relevance was checked, for the simple reason that a close approximation to a real test was aimed for—this type of information would not be available in an archaeological situation.

Selection of attributes

Attributes were chosen for measurement which describe both shape and size. These were considered the best reflection of the similarities and differences between manufacturing styles. Any attributes measuring shape and size were chosen regardless of their known performance at discrimination. Although, theoretically, there are an infinite number of size and shape variables, it was considered a practical sufficiency to measure 24. This is a sufficiently large number of attributes to qualify as a true multi-variate analysis (the Penrose methodology is often performed on as few as six variables).

The attributes (see Figure 15.4 for explanatory diagram)

The following attributes were relevant to Sample One. Note that 24 attributes were recorded for this sample.

Continuous variables were:

1. Maximum length (L1)
2. Maximum thickness (T2)
3. Length of maximum thickness point from cutting edge (L3)
4. Weight (Wt)
6. Width of cutting edge (W1)
7. Width at girth (W2)
8. Width at poll (W3)
9. Index of taper (W1-W3)/L2
10. Index of horizontal symmetry D1/D2
11. Angle of cutting edge
12. High bevel angle (BH)
13. Low bevel angle (BL)

14. Radius of the curvature of the cutting edge (R)
16. Index of width/breadth T2/W2
17. Index of girth position to length L3/L1
18. Index of bevel symmetry (BH-BL)
19. Index of width/length W2/L1
20. Index of thickness/length T2/L1

Discontinuous variables were:

5. Sides: 1 - Tapered. 2 - Rounded. 3 - Flattened. 4 - Dissimilar.
22. Horizontal shape: 1 - Both sides straight. 2 - Both sides concave. 3 - Both sides convex. 4 - Sides dissimilar.
23. Symmetry of cutting edge: 1 - Symmetrical. 2 - Asymmetrical.
24. Presence or absence of chin: 1 - Chin. 2 - No chin.
25. Finish: 1 - Finished well. 2 - Finished roughly.
26. Tilt of poll (reflecting curvature of face): 1 - Poll flat to ground on one face. 2 - Poll never flat to ground.

The following attributes were relevant to sample Two.

Continuous variables were:

1. as before
2. "
3. "
4. "
5. Side thickness (T1)
6. as before
7. "
8. "
9. Index of taper ((W1-W3)/L2)x 100
10. Index of horizontal symmetry ((D1/D2)x 100)-100
11. as before
12. "
13. "
14. "
15 Index of flat-sidedness (T1/T2) x 100
16. Index of curvature of the face ((T2-T1)/W2) x 100
17. Index of girth position to length (L3/L1) x 100
18. as before
19. Index of width/length (W2/L1) x 100
20. Index of thickness/length (T2/L1) x 100

Discontinuous variables were:

21. Relationship of steep bevel (? chin) to horizontal symmetry.
Asymmetrically bevelled: 1 - Chin uppermost; asymmetrical clockwise.
2 - Chin uppermost; asymmetrical anti-clockwise
3 - chin uppermost; symmetrical

Symmetrically bevelled: 4 - Horizontally symmetrical
5 - Horizontally asymmetrical

22. Horizontal shape: 1 - Both sides straight. 2 - Both sides concave. 3 - Both sides convex. 4 - Sides dissimilar.
23. Relationship of cutting edge symmetry to horizontal symmetry. 1 - Edge symmetrical. 2 - Cutting edge skewed to long diagonal. 3 - Cutting edge skewed to short diagonal.
24. Presence or absence of chin: 1 - Chin. 2 - No chin.

It will be noted from the above list that it did not prove possible to measure both samples on identical variables. Although most attributes measured were identical, gross differences between the two groups (particularly the very irregular nature of the lenticular axes from the Eastern Highlands, in contrast to the relatively regular appearance of the planilateral Central Highlands ones) precluded exact duplication. However, every effort was made to use attributes that corresponded closely, in those few cases where identical attributes were not possible.

When all the axes had been measured for these variables, a quantitative comparison of all axes within each sample was performed, using the attributes listed above as the measure of similarity. Calculations were done on a CDC Cyber 76 computer in Fortran IV. Briefly, the comparison of all axes with each other involved standardizing the measurements of the attributes in order to weight them equally, and then calculating the generalized distance between each pair of axes. The measurements of the differences were standarized by using the number of standard deviations by which each axe differed from each other axe on each attribute. This was necessary because it was considered desirable to give equal weight to those attributes whose values have unequal ranges. This worked well for the continuous variables; however, it was not possible for the discontinuous variables. For these variables, a value of distance had to be assigned to 'the same' and 'different'. The two values were counted as being zero for the 'same' and two for 'different'.

The generalized distance between each pair of axes was then calculated, in order to give each pair of axes a numerical value of simllarity; in other words, a measure of the axes' proximity within a 24 dimensional space. The generalized distance between two axes A and B is given by:

$$D = \sqrt{\sum_{i=1}^{n} (A_i - B_i)^2}$$

where n is the number of attributes, $A_i$ is the value of the i th attribute for axe A, and $B_i$ is the value of the i th attribute for axe B. D equals the square root of the sum of the squares of the differences.

The resulting values were converted to a 0-9 scale (in order to facilitate sorting and aid visual display) and printed out in a similarity matrix. The similarity matrix was then sorted in order to place the most similar axes closest together. This is done by continually altering the order, so that the lower numbers are brought closer to the diagonal; i.e. so that axes with low generalized distances are closest together. In this particular research, there is very little ambiguity in interpreting the results: the aim is to see if there is any grouping according to provenance. In many other multivariate analyses, there is guesswork involved in interpreting what the groupings mean. The similarity matrix method was used instead of a dendrogram technique because it gives the full spread of information all the time: it both gives the order and shows how much distance is between each item.

Results

The final sorted similarity matrix for Sample One (the axes that were not traded) is shown in Figure 15.5. The two left-hand columns are the critical ones. Note that axes numbered 20, 3, 19, 27 etc., have been arranged in order of similarity, so that those axes most similar to each other are placed close together. In the far left hand column (the provenance column) axes from the Okapa area are marked +, those from the Arona area are marked X, and those from the Wonenara area are marked O. Note how the Wonenara axes form a very close group, how the Arona axes form a very close group, and how the Okapa axes form a less closely knit group. This is the result expected from the non-traded sample of axes—the axes tend to cluster according to provenance—there are three separate manufacturing styles associated with the three separate areas in which the axes were found.

The final sorted similarity matrix for Sample Two (the axes that were traded) is shown in Figure 15.6. Again, the two left hand columns are the critical ones. Again the axes are arranged in order of similarity (numbers 13, 35, 26, 1 etc.). This time, looking at the far left hand column (the provenance column) note that there is not the marked tendency for axes from the same provenance to cluster that was observed in Figure 15.5. The axes from Wahgi (marked O), from Nebilyer (marked X) and from Kaugel (marked +) do not form discrete clusters at all. This is the result expected from the traded sample of axes—all these axes were made by the group of people who own the Abiamp quarry, so naturally do not cluster in correspondence to their provenance.

Several points should be noted. The bottom six axes in Sample One, and the bottom four axes in Sample Two are clearly odd, as indicated by their high generalized distance from most other axes. In a real test, these axes would have to be excluded from any general conclusions made about the trading patterns of the samples. Likewise, in Sample One, it is doubtful if one could make a definite statement about the Okapa group, because the group is perhaps not cohesive enough, although further trial runs on other samples would help to clarify problems like this.

These results indicate, I believe, that the test has successfully been able to discriminate between two situations: one where the archaeological artefact dispersal was created by free-access exploitation of a raw material, with no trade involved, and one where the archaeological artefact dispersal was created by the artefacts having changed hands between different groups of people; in this case, the transaction being a trading one. The results are clearly promising, and may have important application to situations where trade is suspected; it is certainly a preferable approach to mere speculation. Nevertheless, it would be unwise to register extravagant claims after only one successful feasibility run. There are great complexities relating to trade and exploitation—factors such as the possibility of middlemen, of reworking of the artefacts, of changing styles of manufacturing fashions over time, to mention a few.

| Provenance | Axe Number | Similarity Matrix |
|---|---|---|
| X | 20 | 0 |
| + | 3 | 50 |
| X | 19 | 740 |
| X | 27 | 5440 |
| X | 21 | 64430 |
| + | 6 | 565650 |
| X | 13 | 4444440 |
| X | 14 | 44555440 |
| X | 15 | 445443330 |
| X | 23 | 4554544430 |
| X | 16 | 44534544330 |
| + | 4 | 554444444430 |
| X | 17 | 4544454444430 |
| X | 12 | 45444544434330 |
| X | 26 | 444444444433340 |
| X | 25 | 5444464443333310 |
| + | 46 | 55455654544434320 |
| + | 49 | 654445445444333320 |
| + | 5 | 5544444443443433440 |
| + | 50 | 54544544454444334440 |
| + | 48 | 444444434443433344420 |
| + | 47 | 4554464444444444454320 |
| 0 | 33 | 65564755556555544545440 |
| + | 1 | 444445444444442344444440 |
| X | 18 | 4444444444344434444444430 |
| X | 24 | 44443544444444444544444330 |
| + | 2 | 644444444444444444444554430 |
| 0 | 31 | 5544444344444434444445534330 |
| 0 | 30 | 55443444444344445444446444430 |
| 0 | 36 | 445434344444444445443454434430 |
| 0 | 29 | 5433343444434344443434544443330 |
| 0 | 35 | 46544544444444444544434443434330 |
| 0 | 37 | 445445434444433334422353444443430 |
| 0 | 41 | 4444444344444333444223434444443320 |
| 0 | 45 | 45545554444444444443345444544443310 |
| 0 | 32 | 456444434444544445442454445444432210 |
| 0 | 34 | 6454454444546444554444545454444443420 |
| 0 | 42 | 46544444344443445444346444543333333230 |
| 0 | 28 | 456444344444444454443464445433343443410 |
| 0 | 39 | 4654444344444445444433454454444243324230 |
| 0 | 8 | 66544545445444545444345555544444444442330 |
| 0 | 44 | 466454443444545465554475556544444444432440 |
| 0 | 38 | 5664544443444454444544555665454444445433430 |
| 0 | 43 | 77756445456676778657578777766657656564444340 |
| X | 22 | 578675444556777787674586677766676576754554430 |
| 0 | 40 | 6876656554666565664545766554665454446464565670 |
| 0 | 10 | 76756547666666667566577767777657756775564554590 |
| odd X | 11 | 789796675668988898797798799889899898876774644960 |
| + | 7 | 9976777877878788976979999989896997887677465578480 |
| X | 9 | 69989566576888899989889778777888989897779686598590 |

**Provenance key**

| | |
|---|---|
| + | Okapa |
| X | Arona |
| 0 | Wonenara |

| Provenance | Axe Number | Similarity Matrix |
|---|---|---|
| X | 13 | 0 |
| X | 35 | 40 |
| 0 | 26 | 430 |
| 0 | 1 | 5330 |
| X | 21 | 43320 |
| X | 51 | 433330 |
| 0 | 4 | 4432320 |
| X | 47 | 42332220 |
| X | 56 | 433222220 |
| + | 42 | 4332332110 |
| X | 8 | 43332331210 |
| X | 37 | 433322211110 |
| X | 40 | 4333322122210 |
| + | 38 | 43333332223210 |
| 0 | 57 | 433333322232210 |
| 0 | 72 | 4444433344433330 |
| X | 17 | 44433222233233330 |
| X | 48 | 434432323332233120 |
| + | 16 | 4334313133312222210 |
| X | 14 | 43333122223222132310 |
| X | 19 | 433332322331211212110 |
| X | 18 | 4344423333333323232210 |
| X | 20 | 54444333333332233312110 |
| + | 34 | 434443333333211333231210 |
| + | 22 | 4344433333333123331211110 |
| X | 24 | 42344232333222233322222210 |
| X | 66 | 434543434443333343233322120 |
| X | 15 | 4344433333333333331332222310 |
| X | 39 | 43454342433233333213323221320 |
| 0 | 3 | 544543334333323323232321233310 |
| 0 | 65 | 444443333323323132322212233210 |
| X | 9 | 53444333323332243333322222323220 |
| 0 | 55 | 635443333233333434333233333333210 |
| 0 | 29 | 6454443443334334333432333343332330 |
| 0 | 33 | 65544334444344433434434444554434440 |
| 0 | 60 | 6464434444444444434444344445544343430 |
| 0 | 68 | 5454433344434444333433444454443443330 |
| 0 | 54 | 54443333333334432333334443443334433310 |
| X | 12 | 434433323333333322333333334433233332210 |
| 0 | 10 | 5443333333333332333333333433333333332110 |
| 0 | 6 | 54433333333333332333323333433323333332210 |
| + | 27 | 545443333333333333333322333332122333332220 |
| X | 32 | 5444332332223333132322333343331223233221110 |
| X | 11 | 64544333433344443334434444444433343332322210 |
| X | 59 | 4444333233323333222333333334333333433311222120 |
| 0 | 63 | 6454433343334433233433333443332333333231121210 |
| 0 | 69 | 44443332433333323232333334323233334322222323110 |
| X | 28 | 434443334332333333233333333332333333322222121110 |
| 0 | 70 | 444433343333433332333333333343344333222222312110 |
| 0 | 43 | 43444333433333333213333333331333434433233333121120 |
| 0 | 62 | 645654445444434443343333333333334344433333343321310 |
| 0 | 67 | 5446544344433443333433343333233344443333333322112110 |
| 0 | 46 | 43343332433233333223333333332333444433323333232221310 |
| + | 52 | 43333232332233333322333333333333333434322223332322123310 |
| 0 | 64 | 433333233333333332233333333433333343443322233323233233120 |
| 0 | 7 | 544433333333344333333444444444444444432323333233333433330 |
| X | 45 | 4333232132212333323333333333334334434322223231323134322210 |
| 0 | 41 | 44332322333234333333334444344343444343233333323222243212110. |
| X | 53 | 443323333333344443433334444454444554343333343434333354333310 |
| 0 | 58 | 4444433334433333322223333333333344444333233343333223333322230 |
| 0 | 30 | 545444434433444444444444444454443444434333333233332343334433440 |
| 0 | 61 | 65666444544444443334333333433223334443333234323333333434445440 |
| 0 | 5 | 5455443343333433333333334333333333444434332233323333333433343330 |
| 0 | 44 | 443443434433334333333334434344333444454333344433233243132333334430 |
| 0 | 2 | 534443323223333433333333323323333344433333333333333332224334443330 |
| X | 36 | 656664445554444343344433332323444455544444453433433234444463633440 |
| X | 25 | 6666644465545444444444444444444334444444434434443334333444544545344440 |
| X | 49 | 6566644444444444444444433444434333344444443233334344444445456443353440 |
| X | 31 | 7566655565555555555565444445445444465555454455554545555465665655645650 |
| odd X | 50 | 666555344454444434444434455454444445444334344434444555444444435446554460 |
| 0 | 23 | 87666646666666766666666677776786776665665666666666667766677755686778884 60 |
| 0 | 71 | 788938838838888888883388888888888998898883888888888888888888888888888996980 |

Provenance key

0 Wahgi
**X** **Nebilyer**
\+ Kaugel

## REFERENCES


Adams, R. M., 1974. Anthropological perspectives on ancient trade. Current Anthropology 15(3): 239-258.

Blackwood, B., 1950. The technology of a modern stone age people in New Guinea. Oxford University Press.

Bulmer, S., 1964. Prehistoric stone implements from the New Guinea Highlands. Oceania 34(4): 246-268.

Bulmer, S., 1975. Settlement and economy in prehistoric Papua New Guinea. Journal de la Société des Océanistes 31: 7-75.

Chappell, J., 1966. Stone axe factories in the Highlands of East New Guinea. Proceedings of the Prehistoric Society 32: 96-121.

Hughes, I., 1977. New Guinea stone age trade. Terra Australis 3.

Lampert, R. J., 1972. Hagen axes. Unpublished paper, Department of Prehistory, Australian National University, Canberra.

Renfrew, C., 1972. The emergence of civilisation. Methuen, London.

Strathern, M., 1966. Note on linguistic boundaries and the axe quarries. In: Chappell, J. 1966 q.v.

Strathern, M., 1969. Stone axes and flake tools: evaluations from two New Guinea Highlands societies. Proceedings of the Prehistoric Society 35: 311-329.

www.ingramcontent.com/pod-product-compliance
Ingram Content Group UK Ltd.
Pitfield, Milton Keynes, MK11 3LW, UK
UKHW061847041026
15591UKWH00015B/948